The Library Byelaws and Regulations apply to the loan of this book. It should be returned or renewed on or before the latest date stamped below. Items may be renewed up to three times, by post or telephone at any library, unless reserved by another reader.

PLEASE BRING YOUR TICKET TO THE LIBRARY EVERY TIME YOU BORROW OR RENEW

LIBRARY SUPPORT CENTRE
Ealing Central Sports Ground
Horsenden Lane South
Greenford
UB6 0RP

FINES ARE CHARGED FOR RETENTION BEYOND THIS DATE

LIB.38A 01/00

100402770

Repairing & Restoring Pendulum Clocks

JOHN PLEWES
Certified Master Clockmaker
American Watchmaker's Institute

Sterling Publishing Co. Inc. New York
Distributed in the U.K. by Blandford Press

Dedication

This book is dedicated to my wife, Jackie, for long hours of excellent typing, for many sane suggestions, and for generally keeping my feet on the ground. Without her unfailing encouragement, I doubt that it would ever have seen print.

Drawings by the author.

Photographs by Jacqueline Plewes.

Edited by Frederick Sard

Material in the section "Glass Cutter: Circles" in Part IIII (pp. 172–173) originally appeared in an article by the author in *Workbench* (May–June 1982), and is used herein by permission of *Workbench*.

Library of Congress Cataloging in Publication Data

Plewes, John.
  Repairing and restoring pendulum clocks.
  Bibliography: p.219
  Includes index.
  1. Clocks and watches—Repairing and adjusting.
I. Title.
TS547.P63 1984   681.1′13   83-24262
ISBN 0-8069-5514-7
ISBN 0-8069-7850-3 (pbk.)

Copyright © 1984 by John Plewes
Published by Sterling Publishing Co., Inc.
Two Park Avenue, New York, N.Y. 10016
Distributed in Australia by Oak Tree Press Co., Ltd.
P.O. Box K514 Haymarket, Sydney 2000, N.S.W.
Distributed in the United Kingdom by Blandford Press
Link House, West Street, Poole, Dorset BH15 1LL, England
Distributed in Canada by Oak Tree Press Ltd.
c/o Canadian Manda Group, P.O. Box 920, Station U
Toronto, Ontario, Canada M8Z 5P9
Manufactured in the United States of America
All rights reserved

Contents

	LIST OF ILLUSTRATIONS	7
	LIST OF TABLES	9
	ACKNOWLEDGMENTS	10
	PREFACE	11
PART I	PRELIMINARIES	13
CHAPTER 1	A Word to the Clockwise	13
	Nomenclature: A Break with Tradition	14
	The Number System	14
	Basic Clock Concepts	16
PART II	REPAIRING ANTIQUE-CLOCK MOVEMENTS: AMERICAN, CANADIAN, ENGLISH, FRENCH, AND GERMAN	19
CHAPTER 2	The Basic Repair Text: *American Ogee 30-Hour Weight Movement*	22
	American Clocks	22
	Ogee 30-Hour Weight Movement	22
CHAPTER 3	More American Clocks	33
	Lyre 8-Day Weight Movement	33
	Seth Thomas Regulator No. 2 8-Day Weight Movement	34
	Banjo 8-Day Weight Movement	36
	Mission Longcase 8-Day Weight Movement	38
	8-Day Spring Movement	44
	30-Hour Weight Wooden Movement	49
CHAPTER 4	Canadian Clocks	57
	Pequegnat Regulator No. 1 8-Day Weight Movement	60
CHAPTER 5	English Clocks	62
	8-Day Fusee Movement	66
	Longcase 8-Day Weight Movement	71
CHAPTER 6	French Clocks	84
	8-Day Spring Movement	86
CHAPTER 7	German Clocks	92
	8-Day Spring Movement	94
	Vienna Regulator 8-Day Weight Movement	99

PART III	RESTORING DIALS AND CASES	106
CHAPTER 8	Dials	106
	Longcase White-Dial Repairs	107
	Wooden Dials	110
	Zinc Dials: Stripping and Repainting	110
	Paper Dials	113
	Silvered Dials: Spun-Brass Effect	113
	Arabic Numerals	114
CHAPTER 9	Cases	116
	Longcases	116
	Warps in Longcases	116
	Structural Soundness	120
	Longcase Brasswear	121
	Veneer	121
	School Clocks	122
	Vienna and German Wall Clocks	123
	Surface Treatment of Wooden Cases	124
	Stripping	124
	Staining	125
	Faking Up	125
	Final Finishes	126
	Shellac	127
	Varnish	127
	Wax	127
	Cleaning and Polishing	127
	Glass Domes	128
	Marble or Slate Cases	128
PART IIII	GLOSSARY OF HOROLOGICAL PROCEDURES AND DEVICES	129
	RECOMMENDED READING	219
	INDEX	221

List of Illustrations

Fig. 1.	American Weight Clocks	20–21
Fig. 2.	Ogee 30-Hour Weight Movement	23
Fig. 3.	Pin Removal	24
Fig. 4.	Tightening a Loose Gear	26
Fig. 5.	Resetting a Pin	27
Fig. 6.	Meshing	28
Fig. 7.	Ogee Strike Components	31
Fig. 8.	Lyre 8-Day Weight Movement	33
Fig. 9.	Seth Thomas Regulator No. 2 8-Day Weight Movement	35
Fig. 10.	Banjo Weight Mould	37
Fig. 11.	Mission Longcase 8-Day Weight Movement	39
Fig. 12.	American Mantel-Shelf Clocks	40–41
Fig. 13.	American Wall Clocks	42–43
Fig. 14.	8-Day Spring Movement	44
Fig. 15.	Let-Down Keys	45
Fig. 16.	Winding a Mainspring	46
Fig. 17.	8-Day Spring Strike Components	47
Fig. 18.	Wooden Works Clock	49
Fig. 19.	30-Hour Weight Wooden Movement	49
Fig. 20.	Nylon Bushings for Wooden Plates	51
Fig. 21.	Wooden Works Strike Components	54
Fig. 22.	Canadian Clocks	58–59
Fig. 23.	Pequegnat Regulator No. 1 8-Day Weight Movement	60
Fig. 24.	English Clocks	64–65
Fig. 25.	8-Day Fusee Movement	67
Fig. 26.	Longcase 8-Day Weight Movement	71
Fig. 27.	French Clocks	84–85
Fig. 28.	8-Day Spring Movement	87
Fig. 29.	German Clocks	92–93
Fig. 30.	8-Day Spring Movement	95
Fig. 31.	Vienna Regulator 8-Day Weight Movement	99
Fig. 32.	Straightening a Dial Pillar	107
Fig. 33.	Dial Detail	109
Fig. 34.	Dial Data	111
Fig. 35.	Arabic Numeral Template	114
Fig. 36.	Correcting a Warp, Using a Screw	117
Fig. 37.	Correcting a Warp, Using Saw Cuts	118
Fig. 38.	Correcting a Warp, Using Wedged Saw Cuts	119
Fig. 39.	Saw Cuts in a Warped Door	119
Fig. 40.	Repairing Flat Veneer	122
Fig. 41.	Repairing Curved Veneer	122
Fig. 42.	Correcting a Warp, Using Holes	123
Fig. 43.	Faking Up Blemishes in Wood	125
Fig. 44.	Glass Dome Protection	128
Fig. 45.	Tightening a Loose Barrel Hook	129
Fig. 46.	Bending Tools	130
Fig. 47.	Bushing a Barrel	132
Fig. 48.	Bearing Play	133
Fig. 49.	Bushing a Plate	135
Fig. 50.	An Extended Bushing	136
Fig. 51.	Making Bushings	138
Fig. 52.	Chainmaker	141
Fig. 53.	Wire Mesh Basket	145
Fig. 54.	Perforated Metal Basket	146
Fig. 55.	Cleaning Machine	147
Fig. 56.	Drive Detail	148
Fig. 57.	Clock Stand	151
Fig. 58.	Escapement Types	155
Fig. 59.	Escapement Variants	157
Fig. 60.	Crutch Settings	159
Fig. 61.	Pallet Jig	162
Fig. 62.	Forming a Fly	164
Fig. 63.	English Longcase Fly	165
Fig. 64.	Gear Expander	166
Fig. 65.	Gear Expander—side view	167
Fig. 66.	Upper and Lower Roller Blocks and Mounts	168
Fig. 67.	Top Block, Base and Shafts	169
Fig. 68.	Bearing Block and Mounting Block	170
Fig. 69.	Carrier and Long Bushings	171
Fig. 70.	Glass Cutter: Circles	173
Fig. 71.	Gong Rod	174
Fig. 72.	Hand Repairs	175
Fig. 73.	Tempering	177
Fig. 74.	Hollow Punches	178
Fig. 75.	Huygens System	179
Fig. 76.	Key Shrinker	181
Fig. 77.	Dishing a Barrel	183
Fig. 78.	Measuring a Coiled Mainspring	184

Fig. 79.	Mainspring Repair	185
Fig. 80.	Pendulum Arcs	187
Fig. 81.	Longcase Pendulum Repair	190
Fig. 82.	Pendulum Hooks	192
Fig. 83.	Fitting a New Pinion Leaf	193
Fig. 84.	Pivots	195
Fig. 85.	Polishing a Pivot	196
Fig. 86.	Facing Off an Arbor	196
Fig. 87.	Chamferer	197
Fig. 88.	Repivoting Insert	197
Fig. 89.	Repivoting Steady	198
Fig. 90.	Repivoting an Arbor	199
Fig. 91.	Cap Pivot	199
Fig. 92.	Arbor	200
Fig. 93.	Plewes Chuck	200
Fig. 94.	Chuck Mounted on Lathe	201
Fig. 95.	Chuck Center Shaft	201
Fig. 96.	Chuck Assembly	202
Fig. 97.	Cylinder	203
Fig. 98.	Straps	203
Fig. 99.	Studs	203
Fig. 100.	Clamp Bars	204
Fig. 101.	Hemispherical Jaws	204
Fig. 102.	Knurled Discs	205
Fig. 103.	Chuck Spring	205
Fig. 104.	Stems	206
Fig. 105.	Center-Shaft Front Bearing	206
Fig. 106.	Center Shaft	206
Fig. 107.	Collet	207
Fig. 108.	Center-Shaft Spring	207
Fig. 109.	Center-Shaft Rear Bearing	207
Fig. 110.	Mandrel	208
Fig. 111.	Split Stake	211
Fig. 112.	Brass Gear-Tooth Replacement	212
Fig. 113.	Going-Barrel Tooth Replacement	213
Fig. 114.	Wooden Gear Tooth Replacement	213
Fig. 115.	Verge Escapement	214
Fig. 116.	Brass-Shelled Lead Weight	215

List of Tables

Table 1.	Abbreviations Used in This Book	15
Table 2.	Clock Nomenclature: Traditional and Number Systems	16
Table 3.	Drill Sizes	152
Table 4.	Tapping Drill Sizes	153
Table 5.	Temperature Colors of Steel	176
Table 6.	Pendulum Data	189
Table 7.	American-British Screw Data	218

Acknowledgments

I wish to convey my sincere thanks to Bob, Gail, Gladys, Jane, John, Maurice, and Roy for their freely given advice and assistance in the areas of history and photography. I especially thank Paul who, with a discerning eye, carefully and constructively examined the manuscript and its drawings. I am also much indebted to Art, for his kind help and advice in graphic illustration and other such matters.

Preface

This book is an attempt to fill the need for a systematic, practical bench manual concerning the complete overhaul and restoration of the movements, dials, and cases of old pendulum clocks—those dating from about 1700 to 1940. For beginners, there are step-by-step instructions; for the more advanced, there are interesting machines to make, such as a hand-operated chain-making machine and a special chuck to handle awkward arbor repairs.

This is a hands-on book, in which innovative engineering is favored over old, traditional approaches. Being an evening-course instructor, I have done my best herein to answer the questions students most frequently ask. It is my hope that this approach, along with the step-by-step layout of the repair texts, will allow the book to be of use in technical education courses. I have also included a little history—and some useful, work-oriented theory.

The imperial system of measurement—fractions and decimal inches—has been used throughout, as it would be impractical to do otherwise in a book intended for English-speaking countries. (For example, the great majority of home lathes are graduated in inches.) In all dimensional drawings, typical engineering tolerances are assumed; the more decimal places given for a dimension, the tighter the tolerance required.

Let us assume that you have acquired a clock and wish to fix it yourself, rather than undertake a long and dubious search for a member of that increasingly endangered species—a competent clock repairer.

Before starting any practical work, however, a thorough, cover-to-cover study of this book is strongly recommended. A better overall picture will be obtained thereby, and certain mistakes will be avoided. (For example, read **Drilling** in Part IIII *before* drilling any holes!) As one wag has very tidily put it, "When all else fails, read the instructions!" A browse through Part II will probably uncover a picture of your clock or one close to it, be it American, Canadian, English, French, or German. A diagram of its typical movement will appear nearby, along with the pertinent repair text.

Horological terms and names in bold face in Part II are explained in Part IIII, which is arranged alphabetically. Thus if a **cleaning machine** is required, instructions on how to make one will be found under that heading in Part IIII. This system avoids cluttering the step-by-step repair texts with procedures that apply to clocks in general. Part IIII also explains many clock-related topics not mentioned in Part II.

Part III is devoted to the restoration of clock dials and cases. The former requires the skills of an artist-draftsman, and, as the dial is the most looked-at part of a clock, it deserves serious attention. The latter entails restoring them to a structurally sound and aesthetically pleasing condition. Corrective measures are given for warps and splits, and veneers and finishes are dealt with.

If practical horology is new to you, perhaps a few words of advice may save you much woe: Concentrate on the immediate work at hand, and do it well, before proceeding to the next stage. Never take a short cut by omitting a step—sooner or later, such an omission will return to roost. Always strive for excellence in your work, be it for yourself or someone else.

I

PRELIMINARIES

CHAPTER

1

A Word to the Clockwise

A very prevalent misconception is that clocks are solid objects—like, say, turnips. People who should know better, auctioneers in particular, pick them up and wave them about as though conducting an orchestra. They obviously do not have the tedious job of replacing the pendulum suspension spring and retiming the clock, when they have finished exercising. Nor can they or anyone else repair that old original glass tablet which was cracked by the pendulum bob during its wild gyrations. Free-swinging weights will do even more damage.

Such ruination is difficult, even impossible, to repair—yet easy to avoid. Therefore, before moving any clock in any way whatever, be it two feet or two miles, you should *always remove the pendulum and take off the weights.* Regrettably, many people would still ignore this advice even if it were blazing on their walls in letters of fire, or chiselled onto the gravestones of all clock collectors whose own hands have finally stopped.

As for turning the clock backwards, it depends on the clock involved. In some cases in this book I recommend synchronizing the strike with the indicated time by turning the minute hand backwards. It is also acceptable in a simple timepiece with no striking components to consider. In other clocks damage will inevitably result, except when the hands are moved no more than a very few minutes, and then only on certain parts of the dial. It is really a matter of understanding the construction of various movements. But in general, *do not turn the hands backwards.*

As these two points invariably arise on first acquaintance with a clock, and are responsible for much damage, they are emphasized here at the outset.

Nomenclature: A Break with Tradition

In virtually all clock circles, including the trade, clock gears are known as *wheels*. People new to clocks naturally find this a pointless aberration, which it undoubtedly is. As the term is generally used, it is absolutely incorrect nomenclature: A wheel has no teeth, whereas teeth—which a clock "wheel" unquestionably has—are a gear's reason for being. And the word *gear* is universally understood in all branches of engineering.

During a lifelong association with craftsmen, machinists, draftsmen, designers, engineers, and scientists, I cannot recall gears being called anything but gears. As this seems fair and reasonable, I have had the temerity to continue calling these same devices *gears* throughout this book. I have no quarrel with tradition as such; I simply do not wish to perpetuate an obvious misnomer, however old and venerable it may be. This policy may provoke some criticism, but I feel that it is time to put aside outdated and ambiguous terms in favor of a simpler system of reference.

This is not always an easy thing to do. Clock movements differ considerably in layout and design, and different people use different names for the same component. Nor are American and English terms always in agreement. Having perused a great deal of horological literature, I can only conclude that, despite some agreement regarding nomenclature, no common system, rational or otherwise, exists.

A few examples will suffice to show some of the variations and quirks. *Great wheel = main-wheel = pin wheel = hammer wheel; flirt = lifting piece = lifter;* and *fly = fan = vane = governor.* (*Windfang* is a marvellous name the Germans use for the latter item.)

For a general introduction to the functions of these and other clock parts, see "Basic Clock Concepts," below.

Component nomenclature in the region of the hand arbors can hardly fail to confound the newcomer. The minute-arbor assembly of most American clocks comprises a *center wheel*, a *cannon pinion*, and an *hour wheel*; the last-mentioned pair mesh with a *motion wheel* and *motion pinion*. The minute-arbor assembly of the English longcase, however, has an additional pinion, since, unlike the American design, the minute arbor is part of the actual drive train. In the English movement, the cannon pinion is enlarged and is then called a *cannon wheel* or *motion wheel*. However, E. L. Edwards, for example, in his standard work *The Grandfather Clock,* prefers the term *minute wheel* for both cannon pinion and minute wheel—on the grounds that both rotate once an hour.

Because movements vary somewhat in design, and different people use different names for the same terms, confusion is inevitable. Without a doubt, clocks have far more "wheels" than the ones that rotate!

The Number System

It appears to me that there is a simple way to enumerate the components of a clock train—that is, those parts whose removal would cause the clock mechanism to stop running—be it the time (T), strike (S), or chime (CH) train. Let us start with the source of power and finish with the escapement or fly (whichever pertains). Generally, from three to six arbors are involved in each train, so let us simply call them by number. For example, the no. 1 strike and time arbors will then be *S1* and *T1* respectively, and the gears upon them the *S1 gear* and *T1 gear*. These arbors also carry ratchet gears and main springs (or cable drums) which will be similarly named.

The S1 gear drives the S2 pinion, which is on the S2 arbor. The S2 gear on the S2 arbor then drives the S3 pinion on the S3 arbor, and so on. Besides carrying gears and pinions, the S arbors also carry slotted cams, star gears, or pins, and so on; these can be similarly named.

We thus have a simple system of referring to all the clock components by letters and numbers, instead of by ambiguous and varying names—as far as the actual trains are concerned. This has not involved any radical change; for example, the terms *3d* and *4th wheel* of the old nomenclature naturally suggest what we will call no. 1 and no. 2 gears. It should be noted that changes in function do not change our number system designations. The S1 gear is always just that, for example, whereas under the old and confusing scheme if the *strike great wheel* happens to have hammer lifting pins, it becomes a *pin wheel* or *hammer wheel*.

We still have to account for those gears and pinions that are not actually part of the train itself. The name *cannon pinion* is fairly well known, so we can continue to use it; however, the other non-drive train components must be renamed for clarification. Thus *hour wheel* becomes *hour gear*. *Minute wheel* becomes *reduction gear* and *minute pinion* becomes *reduction pinion* because, in conjunction with the cannon pinion and hour gear, they reduce (by 12:1) the minute arbor's hourly revolution to the hour gear's two turns per day. *Slotted cam* is better than *locking cam*, because this piece does not always perform the actual locking. To some people *locking wheel* equals *count wheel*. One type of count wheel has hour slots but no gear teeth. This logically remains a count wheel. But a count wheel with a combination of hour slots and gear teeth (as in the American "Ogee") then logically becomes a *count gear*. In American clocks *center gear* or *center pinion* is better than *minute-arbor pinion*, which could be mistaken for the cannon pinion.

The number system also enables the pivot holes to be easily and accurately designated. Hence in a repair schedule the notation *S3F* (Strike, no. 3, Front) would refer to the no. 3 arbor's pivot hole in the front plate, all in the strike train. *S3F* is the equivalent of a whole sentence in the old nomenclature; it is easily written, and is much preferable to marking the clock plates.

No great issue is involved in naming the

Table 1. Abbreviations Used in This Book

APX	Approximately
BA	British Association (screw). BA screw nos. are not related to U.S. screw numbers or gauges in any way.
Dia.	Diameter
G.R.O.	Good running order
H.S.S.	High-speed steel
I.D.	Inside diameter
NC	National, coarse (American). "2—56 NC" means "no. 2 gauge, 56 threads per inch, National, coarse."
NF	National, fine (American)
O.D.	Outside diameter
O.G.	Ogee clock
PL	Places. Used in illustrations; means item occurs in other places as well, as in "CHOP 2 PL" (Chop is in two places).
R & A	Retard-and-advance (type of pendulum)
S1, S2, etc.	Strike-train arbor no. 1, no. 2
S1B, S2B, etc	Strike-train arbor no. 1, no. 2, etc. back plate (pivot-hole designation)
S1F, S2F, etc.	Strike-train arbor no. 1, no. 2, etc. front plate (pivot-hole designation)
T1, T2, etc.	Time-train arbor no. 1, no. 2
T1B, T2B, etc.	Time-train arbor no. 1, no. 2, etc. back plate (pivot-hole designation)
T1F, T2F, etc.	Time-train arbor no. 1, no. 2, etc. front plate (pivot-hole designation)
T & A	Time and alarm
T & C	Time and calendar
T & S	Time and strike
T.O.	Time only
TS & A	Time, strike and alarm
TS & Cal	Time, strike and calendar
TS & CH	Time, strike and chime
Thous."	Thousandths of an inch
TPI	Threads per inch
TYP	Typical. Used in illustrations.

Table 2. Clock Nomenclature: Traditional and Number Systems

Typical Traditional Name	Number System Name
Cam, locking cam	Slotted cam
Center wheel	Center gear
Count wheel (without teeth)	Count wheel
Count wheel (with teeth)	Count gear
Escape wheel	Escape gear
Great wheel, main wheel	T1 gear
Great wheel, main wheel, pin wheel, hammer wheel	S1 gear
Hour wheel	Hour gear
Motion wheel	Reduction gear
Second wheel	T2 gear
Warning wheel (O.G.)	S2 gear
Fly, fan	Fly
Motion pinion	Reduction pinion
Unnamed as an assembly	Lifter no. 1
Unnamed as an assembly	Lifter no. 2
Cam-locking hook	unchanged
Count hook	,,
Lifting hook	,,
Unlocking hook	,,
Warning hook	,,
Unlocking pin	,,
Warning pin	,,
Cannon pinion	,,
Crutch wire	,,
Escape cock	,,
Pallets	,,
Pillars	,,
Plates	,,
Saddle	,,
Suspension spring	,,
Verge	,,

parts of a clock, of course; any system may be used as long as a clear understanding of specific functions is maintained. In this book, changes due to the number system have been held to a minimum and involve only those parts that have had several ambiguous names. Of course, in obtaining parts the old terms must be used.

Although the "Establishment" is probably unwilling after so many years to adopt any new system, beginners and students almost always find the old terms illogical and troublesome to learn. A large majority of them ask for a simpler system of reference, so the number system described above seems amply justified on these grounds alone.

Table 1 gives the abbreviations used throughout this book. Table 2 shows how the number system corresponds to the old terminology.

Basic Clock Concepts

The term *rebuilding*, whenever it is used in this book, signifies total restoration of the entire clock to a state as close to its original, factory-new condition as is reasonably possible. It is important to clarify this point, as the word *original* means to some people what *as is* means to others—i.e., with all the abuses of time fully displayed. I feel that the patina of age on a clock case should be preserved wherever possible, but not at the cost of preserving ingrained dirt, open splits, and missing veneer. Likewise, cracked glass tablets and scarred, illegible dials have no beauty. Neglect does not acquire virtue with age.

The knowledge required to complete such a rebuilding embraces at least three distinct separate trades or disciplines: For the clock movement, a clockmaker is necessary; for the dial, an artist-draftsman; while the case is in the realm of the cabinetmaker. It is not realistic to think of clock restoration in terms of the movement alone.

A fourth area of knowledge permeates the other three; it is **history***—the story of several centuries of horological development. As this book, however, is primarily a bench manual, only a brief list of the main historical turning points is given in Part IIII.

Before repairs can be considered, it is most

---

* All terms in bold face are entries in Part IIII, the Glossary, where they are described in detail.

important to be sure of what a clock is and does, how it works, and why it stops working.

In the old clocks currently under consideration, the motive power is derived from **mainsprings,** falling **weights** which pull cables over cylindrical drums, or, more rarely, electricity and a **remontoire.** Energy is stored in these devices by the action of winding the clock periodically. The governor, which regulates the running speed of the clock, is the pendulum. This is coupled to the clock **geartrain** via a device known as an **escapement.** There are several types of escapement, but only the three most commonly found are dealt with in this book. They are the **recoil,** the **Graham** or **deadbeat,** and the **Brocot.** They are used, very roughly, in 75 percent, 15 percent and 4 percent, respectively, of all old domestic pendulum clocks, the remaining 6 percent being made up of relatively rare escapements such as the "grasshopper," the "pinwheel," "Galileo's," and so on. No great accuracy is claimed for these estimates; any census taken will vary from place to place and time to time.

Clocks have up to three **trains** or **geartrains.** A one-train, time only (*TO*) clock is known simply as a *timepiece,* but this terms seems to be obsolescent. (The current word *clock*—in French *cloche* and German *glocke*—originally meant a bell, so that, speaking strictly, a clock is a noisemaker of some kind.) A two-train clock, one with two weights or two winding holes in the dial, has both time and strike (*T & S*) trains. It is called a **striking clock.** A clock having three trains—with three holes in the dial or three weights—is called a **chiming clock,** and has time, strike, and chime (*T,S & CH*) trains. *T,S & CAL* means time, strike, and calendar; and so on. Such abbreviations are commonly used in clock-collecting circles.*

The power of the mainspring, or weight, is applied to the escapement through the train of gears (*wheels* in traditional nomenclature),

---

* They are listed, along with other abbreviations used in this book, in the table of abbreviations on p. 15.

which are usually from three to five in number. The no. 1 gear turns, in an American clock, about once a day, with considerable force behind it. At the other end of the train, the escape **arbor** may rotate more than two thousand times a day, with much less force. Unlike gear trains in most machinery, clock time trains do not run with smooth continuity. They run much more slowly and, due to the action of the escapement, in a series of small jerks. Strike and chime trains, stationary most of the time, run smoothly for short periods when triggered by the time train. This triggering is initiated by the minute arbor, which is either driven from the no. 1 or no. 2 gear, as in most American clocks, or is actually part of the train itself, as in most English and German clocks. The hour arbor is driven from the minute arbor via twelve-to-one reduction gearing, or motion work, as it is known. If a calendar hand is required, additional reduction gearing is added to drive it.

In the strike train, similar forces but higher rotational speeds prevail; the hammer which strikes the hours is driven from the no. 1, 2, or 3 arbor, depending on the type of movement. The fly arbor occupies a similar position in the strike train to that of the escape arbor in the time train. Other arbors, which carry triggering devices known as lifting pieces, or **lifters,** complete the count of devices between the plates of the average clock.

Napoleon is said to have asked the great French horologist Breguet why watches stopped from time to time. "Give me a perfect oil, and I will give you a perfect watch," he was told. The same problem afflicts clocks: Oil makes them run—and oil makes them stop. Depending upon a clock's environment—over a refrigerator, a hot air vent, a fireplace, or in a cool basement—the oil will dry up in roughly one to five years. It will slowly turn to a hard, dry, varnish-like consistency, given enough time. This dried oil is black and firm. It is also impregnated with tiny particles of brass and steel worn away from the clock's plates and pivots, respectively. It thus makes a fairly effective grinding paste, and if the clock continues to run in this condition, as it sometimes does, considerable wear results.

When the clock eventually stops, it does so because of the increased friction caused by this buildup. A similar process takes place in the mainsprings; the coils no longer slide over each other as before, and much less power is delivered. It is therefore usually futile to attribute the clock's stoppage to one single location, as so many do. The clock did indeed stop for one cause—dried-up oil. However, the trouble is not localized in a single place; it is manifested throughout the entire movement, and is invariably accompanied by various degrees of wear.

After long use, besides worn, oval pivot holes and scored pivots one finds the **pallets** in the escapement pitted and needing resurfacing and adjusting. Other faults, too, will be evident. A complete overhaul is indicated by the time a clock reaches this state.

Although the foregoing cycle in the life of a clock is somewhat simplified, it is basically what happens to all clocks every few years. Sometimes a clock that has been abandoned for several years may rust and require more extensive repairs.

The general approach to repairs, then, is to forget the single-fault concept entirely. One simply dismantles the whole movement and examines every last part minutely, correcting every error along the way. Then look for more trouble; it will be there.

The following Part will show how this is done, in thirteen sections with an American movement as our first choice. This sequence is chosen, not because of alphabetical convenience, but because this American type of movement is the easiest to repair. Incidentally, your clock will in all likelihood have seen about a century of hard life at the hands of its various owners, and will doubtless have heard much bad language. It does not need to speak out: The crimes inflicted upon it will be plain to see.

II

REPAIRING ANTIQUE-CLOCK MOVEMENTS:

American, Canadian, English, French, and German

In the repair chapters which follow, the figures given under each movement heading are average values—i.e., those most usually found. It is understood that variations will occur. Pendulum lengths are not given in instances where wide variation is possible.

To avoid repetition, techniques common to all clocks are described in full detail only once, either in the basic ogee repair text (Chapter 2) or in Part IIII. For example, where the fully explained steps of the ogee repair text reappear in later repair texts, they are referred to only by number; they are not described again in full. Subjects such as bushing and pivoting are likewise common to all clocks, but are too basic to be included in any one repair text; they therefore appear in Part IIII for easy reference whenever needed.

When a break occurs in the sequence of step numbers, the missing step simply does not apply to the movement under consideration. As an example, if Step 11 is omitted, proceed from Step 10 to Step 12. However, later repair texts are not simply lists of step numbers with a few gaps. Some steps of the basic ogee repair text are modified in later repair texts to fit the requirements of a particular movement. In these instances, refer to both texts; it will be clear what aspects of the ogee text still apply.

When an imaginary clock dial is used in an explanation, it is to be taken as being observed from the front of the movement, with twelve o'clock at the top.

As clocks differ considerably in design and construction, it is necessary to study all the repair chapters to obtain a complete picture of repair techniques.

Fig. 1. American Weight Clocks

    a. Ogee

    b. Half Column

    c. Four Pillar

    d. Empire

    e. Seth Thomas Regulator No. 2

    f. Banjo

    g. Mission Longcase

CHAPTER

2

The Basic Repair Text: American Ogee 30-Hour Weight Movement

American Clocks

Figs. 1, 12, 13, and 18 show examples of the more usual American clocks, which were made in considerable numbers from about 1790 to 1930. Although variations in case design are legion, the movements inside them fall into only three main categories as far as repair is concerned. There are, first, weight-driven metal movements; second, spring-driven metal movements; and, third, the weight-driven "wooden works" type.

Fig. 1 depicts weight-driven clocks with 30-hour and 8-day movements. The clocks shown in Figs. 1e and 1f are of a somewhat superior quality.

Figs. 12 and 13 show some examples of the more numerous and commonly found American clocks. All are spring-driven and have typical, mass-produced movements of rugged, dependable, everyday quality. Many are 30-hour clocks; probably even more of them run for 8 days, while a few run for longer periods, nominally 14 and 31 days.

Fig. 18 shows an earlier type of American clock, dating from about 1810 to 1845, known as the "wooden works" clock because its movement is made of wood instead of brass. Most of this type run for 30 hours, although some run for 8 days. All are weight driven, and generally T & S.

OGEE 30-HOUR WEIGHT MOVEMENT

Movement: Fig. 2	Cases: Fig. 1a, b, c
Crank size: 4	Weights: 2½ & 3¼ lbs
Pendulum length: 16½"	Bob weight: 1½ oz

As the ogee (or O.G.) is probably the simplest of the American two-train clocks, it is a desirable choice for illustrating general repair procedures.

Ogees were made in several sizes, the smaller ones being the least common. The clock movement under consideration is housed in a 26" × 16" × 4" case, the most usual size. The 8-day version is larger.

Whatever their size, all take their name from the ogee (S-shaped) section of moulding which forms the front of the case. The large, full-length door, which should bear a reverse painting on glass in its lower half, gives ready access to the pendulum, weights, and dial. The two weights are not alike; the 2½-lb. weight hangs on the left, where it operates the strike train, while the 3¼-lb weight drives the time train on the right side. Should the weights be reversed, the clock will strike too fast and run with much less enthusiasm than normally.

As mentioned earlier, the pendulum and

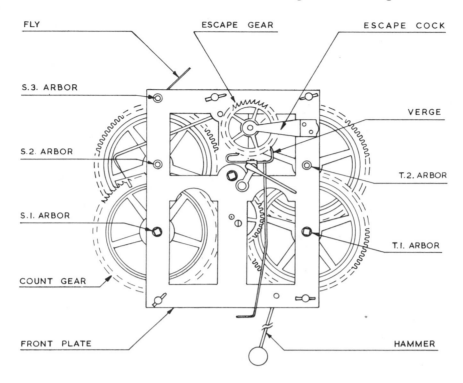

Fig. 2. Ogee 30-Hour Weight Movement

weights should be removed before the clock is set down on the bench on its back for overhaul. Assuming this has been done, you can now open the door for business. (If either weight is wound up too high to be unhooked, it can be removed later, in Step 4.)

Step 1 is to remove the **hands** of the clock. Most people apply pliers to the tapered pin (or nail, sewing pin, tack, etc.) and attempt to haul it out. This is hard on the movement; a much better way is shown in Fig. 3: A gentle squeeze with large pliers is all that is required, and the same method is used to replace the pin. This idea is applicable to all other clocks that use this hand-retaining configuration.

As the various parts are removed, put them into small, shallow, open dishes—say, 5″ across and 1½″ deep. Never set them down on the bench or they will get lost or damaged. It is best to use two such dishes: no. 1 for items destined for the **cleaning machine** and no. 2 for everything else.

The condition of the hands dictates which dish will receive them. If they clean up fairly well with a brisk rub and a little metal polish, and the **bluing** is still good, use dish no. 2. If they are badly pitted with rust, use dish no. 1, and see step 25 (p. 28) for further details.

The hour hand is taken off using an upwards rotating pull on its center boss—not on the hand itself, lest it break off.

Step 2. Remove the dial by turning the two bent, headless nails that secure it along its bottom edge and lowering it out from under the similar pair of nails that hold it at the top.

Step 3. The pendulum **suspension spring** can now be gently eased out of its slit in the pendulum cock, and slid out through the elongated loop at the end of the **crutch** or **verge wire.** Care must be taken during this operation

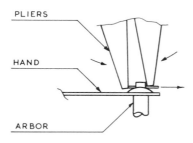

Fig. 3. Pin Removal

to see that the suspension spring is not deformed, even slightly.

Step 4. If, as previously mentioned, the weights are wound up high alongside the movement, then they are next to be removed. As their cables are generally in need of renewal, it is simplest to cut them, keeping the pieces as a length check for new ones. If the cables are to be used again, and are difficult to unhook, then the **clicks** of the ratchet gears on the winding drums must be released with a scriber or small screwdriver. Holding off the clicks thus will enable the weights to be slid down to the bottom of the case and unhooked. Then untie the knots and pull the cables out of their holes in the winding drums. If the cables are weak or worn, cut them and discard them.

Step 5 is to remove the movement. Unscrew the mounting block at the top of the movement. If it is not free to slide out, check the ends of the seatboard to which the movement is hook-bolted. Two holes of some ¾″ diameter are to be seen, and protruding into them two small nails. These prevent the seatboard from sliding out of the slots in the long ribs which support it between them.

Removing these with long-nosed pliers enables the movement and its seatboard to be withdrawn. The hook bolts can now be loosened and the seatboard removed, wiped clean, and deposited in dish no. 2. The movement is now ready for inspection.

Step 6. If the movement is well coated with dirt and liquid oil, it is a good idea to brush this off, using a general purpose solvent. As these solvents leave a thin film behind them, they must be used only before the clock is cleaned, never afterwards. They are of use only as a primary rinse for excessively dirty clocks. Use of such a solvent instead of the stronger and more costly solution in the cleaning machine is to avoid unduly polluting the latter. General purpose solvents should otherwise be kept away from clock mechanisms.

Step 7. Look over the movement to see that it is complete and to find any excessive wear and serious damage. The component most usually missing is the **verge** (see Fig. 2), for it is possible for it to fall off and become lost. (See under **verge** in Part IIII if a new one is required.) For those who doubt their ability to reassemble the "works," now is the best time to make notes or sketches for later use.

Ogee movements that have more than three arbors in each train are quite rare, so we shall consider our movement to be of the usual three-arbor-per-train type. With two trains and five other arbors not part of any train this remarkably simple and reliable design has a total of eleven arbors. These five other arbors carry two lifters, the strike hammer, the minute hand, the **hour pipe,** and the reduction gear and pinion.

The ends of all clock arbors are reduced in diameter to form **pivots** which pass through pivot holes in the two clock plates. As previously mentioned, wear is most evident at these points, often being indicated by the presence of small black deposits. When these are gently removed, a surprisingly large hole in the brass is sometimes revealed. Experience is required to know when a pivot is loose enough to require **bushing,** the escapement being the most critical area. A common configuration is for an arbor to have an integral gear-and-pinion combination near one end. The bearing play at that end is then more critical than at the other end, as it affects the **meshing** of the gear train to a greater degree. (See Fig. 6, p. 28.)

In general, a pivot requires bushing when it has more than about 0.005″ play. (As a guide to beginners, a single human hair is about 0.002″—two thousandths of an inch—in diameter.) Any pivot hole that has suffered the attentions of a center-punch artist must always be bushed. Vandals who punch a ring of pits around a pivot hole in an attempt to close it up are much too common. The use of hole-closing circular punches is not much less reprehensible, for both these destructive, short-term makeshifts result in damaged plates and scored pivots. They render more difficult a sound repair job. Not only is bushing required, but the pivots will then have to be machined true, instead of simply polished. Such damage is to be found on both sides of both clock plates. (The full details of pivot fits and bushing are given in Part IIII.)

Step 8. Make a list of all the pivot holes that obviously need attention, bearing in mind that others may come to light when the movement is taken apart.

The plates of most old clocks are generally somewhat scored, and many repairmen mark tiny crosses at holes that they intend to bush, on the grounds that one more scratch will hardly show. As such crosses are almost filed out when bushing is complete, this practice causes no great damage. It should never be used on fine clocks, however, as a decrease in value would result. Undoubtedly the best approach is to write down the positions of worn pivots. While this is somewhat awkward in the old nomenclature, the number system is very easy to use (explained on p. 14).

Step 9. Make a rough drawing of the positions of the **warning** pins as they lie when striking is over; some clocks have two gears to carry such pins.

In all striking and chiming trains, it is not enough to replace the various arbors in their correct, original positions; they also must be "phased," as it were. In the ogee, a small L-shaped wire on the fly arbor, S3, constitutes the warning pin (see Fig. 7f). The position at which this pin stops when the clock has finished striking should be noted. At this point the strike hammer has to be entirely at rest; it must not be raised at all, even slightly. This means that none of the hammer-lifting pins on the S1 gear contact the hammer-lifting tail until striking actually commences.

Unless these conditions exist, the strike train is incorrectly set up in any case. Failure to note these points will doubtless give rise upon reassembly to the hard words previously mentioned, because the strike train will malfunction.

Time-train arbors and gears are straightforward and unencumbered in appearance, whereas those of the strike train have miscellaneous slotted cams, warning pins, star gears, and so on attached to them. Assembly will be easier if this is remembered. Another point is that the larger and sturdier the arbor, gear, or pivot, the nearer it will be to arbor no. 1, the source of power.

A further item which can bring trouble is the **count wheel,** or **count gear,** as I prefer to call it. This gear is mounted on a boss which surrounds the S1F (sometimes S1B) pivot hole, and is driven by a pin on the S2 arbor. It is all too easy to reassemble the count gear back-to-front on its boss; this will make the clock strike backwards—i.e., 9:00, 8:00, 7:00, etc. Although this is an interesting enough phenomenon, it is hardly useful and its correction entails taking the clock apart again. As time and trouble have usually been taken in setting up the strike, repeating the process will go unappreciated. In common with many other large American clock gears, the count gear often has a pressed-out channel around its periphery. The convex side of this channel almost always faces forwards—i.e., towards the front plate. It indicates its own correct assembly—if noted before dismantling.

Step 10, then, entails marking not only this gear but also a few others, to avoid any confusion during reassembly. A tiny "D" for dial-side or "F" for front should therefore be scribed on the count gear. This is particularly worthwhile when the gear is flat, which in a great many clocks it is.

In an ogee, both the S1 arbors and their gears are quite dissimilar, and can hardly be mistaken for one another, but this is not the case in most other clocks. It is therefore important to adopt a policy of marking the S1 gear with a small "S" in all cases. When the arbor assembly is demountable, it pays to mark each part in a like manner. In a time train of five arbors, nos. 3 and 4 can sometimes trade places, so it is wise to mark one of them with a "3" or "4."

Step 11 does not apply to the ogee (see page 19). Proceed to Step 12.

Step 12. Remove the pillar pins which hold the plates together, using the method shown in Fig. 3. The original square-section tapered pins will very likely have been replaced by tacks, small nails, sewing pins, toothpicks, bent wire, and so on. In many pinning applications, a wire of suitable diameter, properly bent into a tidy Z shape, is superior to a tapered pin, for it cannot fall out. However, since it is not tapered, a wire does not exert a tightening action. Wire would not be suitable for retaining clock hands because of removal difficulties and its unattractive appearance.

We are now ready to dismantle the movement: Gently pry the plates apart with your fingers and then extract the arbors. Slide the hour pipe off the minute arbor.

Step 13 does not apply to the ogee. Proceed to Step 14.

Step 14. The count gear is friction-held on its boss by a Y-shaped brass friction spring. Two arms fit into a groove in the boss, and the third is bent down at its extremity to fit into a small hole in a **crossing** (spoke) of the count gear. Remove the count gear by gently prying up this third arm, or "tail," and slide the friction spring off. Also extract the pin which passes through the rear end of the minute arbor, and disassemble the three-armed friction spring (known as a **center spring**) and then the center gear.

Step 15. Put all the parts from the above dismantling operation—which should have been put in dish no. 1—into the cleaning machine. In most cases, twenty minutes running time is enough to bring these parts up to a bright shine. The results will, however, vary with the type and condition of brass used in the particular clock. Some parts will never attain any great brilliance even after long immersion. (See **cleaning process** in Part III.)

Some people prefer to effect repairs before cleaning. This hardly seems desirable, as dirt or black gum may obscure wear or damage such as a crack in a gear, all of which would require rectification. Soft soldering, for example, cannot be done on dirty metal. Working on clean components is also much less messy and results in greater accuracy.

Step 16. The cleaned, rinsed, and dried clock parts are now individually inspected. Take the gears first: Each tooth must be examined against the light to see that it is symmetrical and not bent or deformed in any way. Missing or broken teeth must be replaced (see **Tooth replacement—gear** in Part IIII). Teeth which are not too badly bent can generally be straightened with a small pair of long-nosed pliers. It is a good idea to grind small flats on the outside of the tips of these pliers, to facilitate their entry between gear teeth. Pliers so

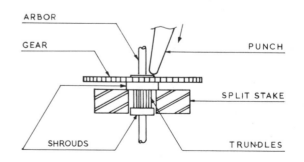

Fig. 4. Tightening a Loose Gear

ground are also useful in removing any faulty **trundles** from **lantern pinions,** of which there are four in an ogee movement (see Fig. 4).

Step 17 does not apply to the ogee. Proceed to Step 18.

Step 18. The brass collets, or shrouds, of lantern pinions are sometimes loose on their arbors. Check all shrouds by firmly gripping each one in turn with the pliers—but not firmly enough to mark them badly—and then testing the gear to see if it is loose enough to turn. This method will tell if the shroud or the gear collet is loose on the arbor. It will also reveal any gear that is loose on its collet.

The gear collet can be hammered tight again, if a **split stake** is available (see Fig. 4), otherwise it is best to soft solder it. Soft soldering is also an excellent method of fixing a loose shroud.

Some would-be purists decry the use of soft solder in clocks altogether. It is enough to point out to them that in a great many English longcase movements both the escape gear collet and the anchor collet have always been soft soldered onto their respective arbors. As these old-timers date from as far back as 1670 or even earlier, this seems precedent enough for anyone.

The **cannon pinion** is a drive fitted onto the minute arbor, and there is no occasion to remove it. Should it be cracked, as it sometimes is, then soft soldering is by far the best remedy, if care is taken that none of it appear on the front side of the pinion to foul the hour pipe.

Step 19. Take the fly off its arbor and remove the dirt from its bearing grooves. The arbor is then cleaned with fine emery cloth or a polishing stick, and replaced in the fly. Judicious bending of the split sections of the fly will ensure a firm, but not tight, fit. The fly's purpose is to govern the speed of the strike train and cushion the impact of the sudden stop which occurs when striking ceases. It therefore must never be fixed rigidly to its arbor, as some people seem to think. In contrast, the L-shaped warning pin on the fly arbor has to be tight in

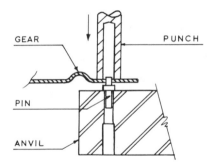

Fig. 5. Resetting a Pin

its hole, as also must the rather similar unlocking pin on the minute arbor.

Step 20. Other pins, such as the hammer-lifting pins, must also be tight and firm in the rim of the S1 gear. Sometimes a few of these pins were not set down far enough originally; this can be corrected by means of a **hollow punch** and an **anvil** (see Fig. 5). After being thus tapped down into position, the end of each pin is then riveted over with a small hammer, in the usual fashion.

Step 21. Check the **pendulum cock** to make sure that the **suspension spring** slit is vertical and that no play is present. If necessary, tighten it in the same way as the hammer-lifting pins.

Step 22. See that the verge pin is tight in its arm; if it is loose, replace it with one of a slightly larger diameter. A tapered steel pin, put in from the rear, may be used, provided the holes in the verge saddle are also broached from the rear to suit it.

Step 23. Test the firmness of the escape-cock rivets, and tighten them if necessary. A little hammering may be necessary to tighten up the four pillars on the back plate, if they have any tendency to turn.

Step 24. Remove any rust from the lifters with a strip of fine emery cloth, clamping the arbor in a vise (with brass chops, to avoid vise marks). Loose hooks are difficult to re-rivet; soft soldering is an easy and adequate cure.

When lifters cannot be readily mounted in a

lathe to have their pivots polished, they must be hand held. A flat emery polishing stick is used in either case. Hand polishing does not guarantee roundness but is sufficiently accurate in this case, because lifters merely rock—they do not rotate as other arbors do.

Step 25. To deal with rusty clock hands, lay them on a wooden block with a 5/16"-diameter hole in it deep enough to take the boss of the hour hand, and clean them with a sanding block and fine emery cloth. This will reduce the rust and bring out the bright steel underneath. It is best to reserve heat **bluing** for smoothly surfaced parts; badly pitted hands do not look well when heat blued. The latter should be covered with a dark, gun-blue paint, if it is available; otherwise black paint may be used.

Step 26. Carefully check the play in the winding clicks; these are on the two no. 1 gears. These clicks are often loose, and must be firmed up so that they mesh properly with their ratchet gears. (See **click** in Part IIII.) The sequence in which you do such jobs as click repair, tooth replacement, and rust removal is not important, but pivots must certainly be dealt with before any bushing is considered.

Step 27. Polish or repair the pivots of all arbors in the manner given under **pivots** in Part IIII. This operation requires a lathe. Even then some arbors are difficult to mount so that good access to their pivots is obtained. When the pivots are sound—and only then—it is time to attend to the bushing or to worn holes.

Step 28. Making and inserting bushings is not difficult, but it requires close attention to proper procedure. See **bushings** in Part IIII before proceeding further.

Step 29. After bushing both clock plates, wherever necessary, clean out all pivot holes with **pegwood;** this process is called **pegging out.** Although it is a simple job, and one common to all clocks, it is sometimes overlooked on reassembly. This means having clean pivots in dirty pivot holes—an unsanitary combination.

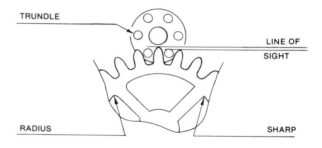

Fig. 6. Meshing

Step 30 does not apply to the ogee. Proceed to Step 31.

Step 31. Test the meshing of all the gears and pinions that are involved with those arbors whose pivots have been bushed. This is done to ensure that the pivot holes have not been displaced from their original positions, and also to see that a good mesh has been obtained. See Fig. 6 and **meshing** in Part IIII. When trying out two or three gears for correct meshing, it is a good idea to check their arbors for end play at the same time.

Step 32. Any warps present in the clock plates may have been deliberately put there by someone to take up end play, or may be accidental. The plates should be kept as flat as is consistent with reasonable end play, which, in American clocks, means 0.010" to 0.020". It is not advisable for the end play to exceed 0.020", although many clocks, such as ogees, have 0.025" or more. In spring-drive clocks, where the pressures are greater, such a large degree of end play could be dangerous in a no. 2 arbor, for example.

Step 33. The center gear and no. 1 gears must now be mounted on their respective arbors and pinned in place. Make sure that the three arms of the friction springs exert a firm, but not excessive, pressure on their respective gears. This is done by slightly bending the ends of the arms, while taking care that the spring is not tilted when the pin is in position.

Step 34. Replace the count gear in its correct position on the front plate, and see that its friction spring exerts only a mild pressure—

just enough to prevent the gear from spinning freely.

Step 35. Inspect the hammer spring, and replace it if it is short or broken. Make sure that the hammer head is tight on its arm. Should the head be loose, soft solder it; hammering it is not effective.

Step 36. Clean out the hour pipe with a cotton swab. Make certain that it rotates freely on the minute arbor at any depth of insertion.

Step 37. Give all parts a final inspection; see that there is no dirt left in the pinions, etc., and then ready them for reassembly.

Step 38. There are some quite elaborate movement-assembling devices on the market, but a simple metal 4″ × 5″ **assembly box,** about 2″ deep, is all that is needed. A circular version is also usable. Even a 20″ metal strip bent to form a rectangle would be adequate, although the box is more useful, for it can store an anvil, etc. The box is open at the top, and is used for supporting the clock plate during reassembly. It raises the movement off the bench, and thus gives clearance to pivots, winding squares, and the like, which protrude underneath, and keeps them clean.

Lay the rear plate of the movement, pillars up, on the assembly box, and place the arbors in their correct positions. Set the hour pipe on the minute arbor, but leave out the escape arbor for the present. If in doubt as to placement, refer to Fig. 5 or to your previous notes made when disassembling the movement (see Steps 7–12).

Movement assembly is usually easier when the lifters are installed before the other arbors, even though they have shorter pivots and often fall out of position. Placement of the lifters sometimes causes a pause in the proceedings. Usually there are two positions available for them, and the question is, which goes in which? The solution is to decide which lifter lifts the other one and what lifts the first lifter. The action starts with the unlocking pin on the minute arbor (Fig. 7a; note that Fig. 7 shows lifters and other components not readily discernible in Fig. 2). The lifter with the long U-shaped arm, the **unlocking hook** (Fig. 7b), is the only one that can be reasonably expected to connect with it. In the ogee movement, this lifter goes into the higher of the positions. A further check is based on the fact that, were the other lifter put in this position, the **count hook** (Fig. 7c) could not possibly contact the teeth of the count gear. Thus, correct placement of the lifters can be ascertained.

With the arbors installed as described, put the escape gear in place under the escape cock, and, holding it thus, carefully lower the front plate over the hand arbors and winding squares. Now set the rear pivot of the escape arbor into its pivot hole in the rear plate. Applying slight downwards pressure to the front plate, move the arbors of the hammer and the reduction gear gently until their pivots enter their respective pivot holes. To maintain the position thus gained, temporarily slip a couple of pins or short wires through the holes in both lower pillars. The remaining pivots are then positioned likewise until the plates can be closed up. It is useful to put temporary pins through the two top pillars also.

Step 39. Check that all arbors have adequate end play and rotate freely. The meshing of the gears and pinions (Fig. 6) must also be tested. Place a finger on both the front plate and escape gear simultaneously, so that rotation cannot occur. When the T2 gear is then twitched with another finger, there should be enough backlash to produce a clicking noise. Repeat this procedure with all gears and pinions, rotating them to ensure that there is backlash in all positions. If such freedom is not universal, a bind will occur which will stop the train at those points where backlash is nonexistent or insufficient.

On the other hand, excessive backlash will stop the clock, as insufficient power will be transmitted. Although essential, backlash must be held to a minimum.

Step 40. Fit a clock hand on the minute arbor and turn it slowly clockwise. Note the action of the lifters, and check that they work in correct sequence. This will bring to light any hook lying on the wrong side of an arbor. When this happens, it is sometimes possible to ease the hook into position, but often it is necessary to take the plates apart to correct matters. This may be tiresome but is much better than breaking a pivot or hook by using undue force.

Step 41 is to set up the strike train. It is interesting to try to set up the slotted cam and the warning pin in their correct positions during initial assembly. However, because the arbors generally rotate a little before meshing occurs, accuracy is seldom achieved; still, an approximation is helpful. After assembly, the strike train usually runs continuously, when the S1 gear is pushed round by hand; this is because the train is unable to lock itself. Further adjustments are then required.

Take out the strike-side temporary pins—i.e., those in the top center and bottom left pillars. Turn the minute arbor until the lifters fall, and see that they do not interfere with subsequent operations; wiring the escape gear is hardly necessary. Now push the S1 gear until the tail of the hammer has just fallen off one of the thirteen lifting pins, and keep it in this position. The slot in the cam on the S2 arbor should now be under the cam-locking hook (see Fig. 7c).

If this is not the case, gently pry the plates just far enough apart to remove the fly and to disengage the S2 pinion from the S1 gear. Rotate the S2 arbor until the slot is under the cam-locking hook and then reengage the pinion with the S2 gear. During extraction and replacement of the S2 arbor, care must be taken to avoid bending the pivots.

Test and repeat this operation until the cam locking hook drops into the slot in the cam on the S2 arbor, on which it rides, and reliable locking is achieved.

The cam-locking hook may be unable to drop because the **count hook** (Fig. 7c)—which is on the same arbor—does not fall between the teeth of the count gear. The end of this hook must then be carefully bent until the flat blade thereon falls centrally in the deep, hour-marking slot in the count gear; Fig. 7d shows the bending points. Fig. 7 also shows the part names of lifters and other components which are not readily discernible in Fig. 2. Centering the count-hook blade is essential for reliable operation. (It cannot be achieved, for example, if the pivot holes of the count hook arbor are too large; this is occasionally overlooked.) When properly adjusted, the blade of the count hook will coincide with a radius of the count gear, as shown in Fig. 7d.

Turn the S1 gear and allow the train to run and lock. With the train in this position, carefully replace the fly, without disturbing the S2 arbor, with the warning pin (Fig. 7f) pointing to the count-hook blade.

The full operation of the strike train can now be tested. Install the verge so that the time train cannot run freely, and set a minute hand in place on the square of the minute arbor. Apply finger pressure to the S1 gear and slowly turn the hand until the cam-locking hook rises, allowing the fan arbor to turn until the warning pin hits the warning hook. This completes the **warning.** If necessary, the hand should now be removed and repositioned to indicate about 5 minutes to the hour.

Check that the hammer is still down; it must not be even slightly raised at this stage. Now continue to turn the hand until striking occurs, followed by secure locking. If the locking is at all doubtful, small adjustments of the lifters may be required.

The difference in the depths of the slots in the count gear enables the cam-locking hook to determine whether or not the train shall run. The angle between the arms of the count hook and the cam-locking hook is critical. It can be adjusted by inserting a scriber or small screwdriver between these two parts to gently lever them farther apart or closer together, as required. This enables the correct depth of locking to be obtained.

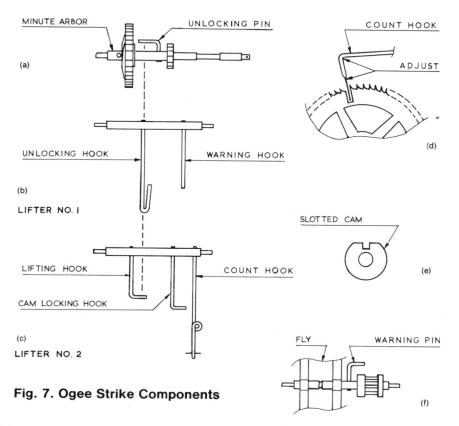

Fig. 7. Ogee Strike Components

Should striking commence too early or too late, the end of the unlocking hook must be raised or lowered relative to the unlocking pin on the minute arbor. It is of no avail to open or close the hook, and only a small adjustment can be obtained by twisting the unlocking pin slightly out of alignment with the minute arbor. No great changes are necessary, but such adjustments must be made carefully.

When all is satisfactory, the four pillars must be secured permanently with tapered pins or short Z-shaped wires, as previously mentioned.

It is not feasible to describe all of the problems that might obviate setting up the strike, but the foregoing portrays the essentials. When the action is understood, the remedies suggest themselves, and a little judicious bending or "gear shifting" usually suffices. Fig. 46 (in Part IIII) shows two tools which are very useful for this purpose; they are easily filed from 0.250" mild steel rod, the business end being tapered down for ease of entry.

Step 42. The verge must now be considered. It is safe to say that most people have more trouble with it than they care to admit. As it involves heat treatment of steel and is too integrated with the escape gear to be considered alone, it is explained in Part IIII under **Escapements** (the "Recoil—Strip Pallet—American" type).

Step 43. If the escapement is correctly set up, the weight of the ogee movement is enough to make both trains "run" (or "run free"). This useful check is done with the pendulum rod off, and the two no. 1 gears held between the thumb and fingers of each hand. A slight tilt to left or right may be necessary to balance the verge, and the count hook must be lifted to allow the strike train to function.

Step 44. Under normal running conditions the time and trouble involved in replacing a clock's relatively short cables is so great—far outweighing their monetary cost—that it is better, unless the cables are quite new, to routinely replace them after an overhaul, when the movement is already out of its case for repair.

The best cable to use for ogee or similar clocks is braided nylon; braided cotton will not wear as well. Braided chalk line or fishing line, of 50–100 lbs breaking strain, is very suitable. Do not use a twisted cable, for the weights will spin and this may cause trouble. Cables of metal, such as bronze or stainless steel, are too stiff and harsh, and will wear the drums.

When the weights are fully down, only one turn should remain on the drums; this gives a tidy winding pattern and prevents excessive layering. When fitting cables, do not leave a tail of cable sticking out from the hole in the drum, as this may disrupt an even winding pattern. The knot should sit flush in the hole provided in the drum, but of course must be large enough not to pull right through.

Step 45. When it is clear that the movement is mechanically sound, it should then be oiled. Refer to **oiling** in Part IIII.

Step 46. Before recasing the movement, check that the two wooden pulleys at the top of the case are free of cracks and chips and spin freely. There is sometimes a third wooden pulley—in the left-hand wooden rib inside the case. This must be likewise checked out; the strike cable passes under it and so avoids fouling the fly.

Step 47. If the two wooden ribs are firm and the case requires no further attention, then it is time to reinstall the movement. The pendulum-rod hook is handy for pulling the cables through the pulley holes. The weights are attached to the free ends of the cables by small S-shaped wire hooks which have one end closed. A useful way of recalling which weight goes where is to remember, "When *might* is *right,* there's *little left* for me." (The "mighty" weight is on the right and the little one on the left.)

Step 48. Check that the strike-setting wire (if present) is in place, and that it passes through the staple in the wooden seatboard. This wire usually hangs from a loop in the count hook with its free end extending below the dial. Make sure that it does not interfere with the counting action of the hook blade. Upwards pressure from a fingertip on the end of this wire raises the count hook and thus actuates the strike train. It thus enables you to synchronize the strike with the time indicated by the hands.

When there is no provision for a strike-setting wire, the clock hand must be used instead. Turn the minute hand clockwise to the 57-minute mark and note that the warning occurs. Now rotate the hand counterclockwise back to about 40 minutes; this will cause the clock to strike as if at the hour. Then move the minute hand up again to the 57-minute mark and repeat this forward and backward cycle until the strike agrees with the time indicated on the dial. If the hour mark is accidentally passed, allow the clock to strike, keep going clockwise, and then continue as before. Do not move the hands backwards once the hour mark is passed, or damage will result.

Step 49. Wind the weights an inch or so off the bottom of the case, and carefully replace the pendulum rod and bob. Do not wind the weights up fully. Set up the beat and adjust the crutch wire; refer to **pendulum** in Part IIII. Start the pendulum swinging, and check that the clock runs. Should it fail to run, the weights can easily be removed for investigation. (This would not be the case had they been wound up fully.)

Step 50. Replace both hands and allow the clock to run a day or two before adding the dial, as a few small adjustments are sometimes required.

CHAPTER

3

More American Clocks

LYRE 8-DAY WEIGHT MOVEMENT

Movement: Fig. 8	Cases: Fig. 1d
Crank size: 5	Weights: 7 & 9 lbs
Pendulum length: 20¼″	Bob weight: 3 oz

This popular clock has a case measuring some 33″ high × 19″ wide × 6″ deep. Rosewood and mahogany veneers are common, with a lesser number of oak.

The movement is larger than the 30-hour ogee type, and the plates are somewhat thicker, being 0.072″ as compared to the usual 0.058″ of most American movements. Hence any bushings required will be some 0.085″ long instead of 0.070″. Five pillars hold the plates together. The no. 1 gears and drums are larger than those of the ogee. Each train has four arbors; the arrangement of the S2, S3, and S4 arbors in the 8-day lyre is very much like that of the S1, S2, and S3 arbors on the 30-hour ogee, and it works the same way.

Apart from size and an extra gear in each train, the differences between the movements are minimal. The plates are lyre shaped instead of square, and the strike-setting wire passes through the center of the movement instead of hanging down the front. Otherwise the 8-day lyre (and the 8-day ogee movement, too) can be looked upon as a 30-hour ogee movement with new winding arbors added below to give 8 days of running time.

The weights are square in cross-section; like the round-sectioned ogee weights, they taper, being widest at the bottom. The lighter (7 lb) weight operates the strike train (left side), as in the ogee. (Tapering weights of similar poundages, but with a rectangular section, are used in the 8-day version of the ogee.)

Repair procedures to the lyre (and to the similar, but rectangular, 8-day ogee movement) are the same as Steps 1–50 of the 30-hour ogee text (Chapter 2).

Fig. 8. Lyre 8-day Weight Movement

SETH THOMAS REGULATOR NO. 2 8-DAY WEIGHT MOVEMENT

Movement: Fig. 9	Case: Fig. 1e
Crank size: 6	Weight: 6 lbs
Pendulum length: 22¼"	Bob weight: 2 lbs

This sound, heavy clock frequently has a case of solid oak, but a walnut-veneered version is also sometimes found; its overall size is 36" × 16" × 6". As the pendulum cannot easily be removed, the rather heavy bob must be taped down or otherwise immobilized before the clock is moved at all. The bob can, of course, be removed from the shaft, but this necessitates rating the clock again from scratch.

As originally suspended, the brass-shelled weight (measuring 6¼" × 1¾"), is also awkward to remove. It is generally necessary to take the pulley apart, which entails unscrewing the pulley arbor. Reproduction replacement weights have a more convenient arrangement using a simple open hook.

At the risk of being repetitive, I stress once again that, be it easy or difficult, both the bob and the weight must be dealt with (*Step 4*) before the clock is transported in any way. The amount of cracked glass found in the doors of these clocks points to the frequent neglect of this precaution.

The lower door opens with the clock's winding crank, and the hands and dial are easily removed. The well made T.O. (time only) movement has trapezoidal plates 0.087" thick; hence, any bushings required will be 0.100" long. The four-arbor train is fastened with four sturdy brass pillars; and **maintaining power** is incorporated for the deadbeat escapement. Four other pillars—of steel—pass through the front plate and secure the back plate to a substantial cast-iron mounting screwed to the back of the case. Unscrewing these four captive pillars enables the movement to be withdrawn, leaving the pendulum and suspension spring still pinned to the mounting bracket.

Braided brass or bronze cable is often found on these clocks, but today's braided nylon is much kinder to the winding drum. Whichever type of cable is used, it must now be removed from the drum and from the stud holding the fixed end. If a sound metal cable is to be used again, it goes into the no. 1 dish and eventually to the cleaning machine.

The no. 2 arbor in the drive train is also the minute arbor, whereas in the ogee the minute arbor is not part of the drive train but is driven from the no. 1 gear separately. The Seth Thomas No. 2 arrangement therefore uses one less gear than the other system; most European clocks use the same idea.

Dismantle the movement and unpin the T1 gear from its arbor (see **clicks** in Part IIII). Pull the minute pipe with its cannon pinion off the minute arbor, and then clean all parts per *Step 15* of the basic ogee repair text.

Continue with *Steps 16, 18, 21, 26, 27, 28, 29, 31, and 32*. An end-play check will show a much tighter movement than in the ogee, the measurements being from 0.005" to 0.010". If these tolerances are not maintained, the escape gear can run partially off the edges of the rather narrow deadbeat pallets, resulting in a most undesirable pattern of wear.

As there is no center gear, *Step 33* applies to the T1 gear, as well as to the clutching action of the minute pipe on the minute (no. 2) arbor. Should the pipe be too loose, a very slight squeezing of the two sides of the cutaway pipe will correct matters.

The small brass bracket on the underside of the movement stops the pulley from entering the movement on completion of winding. Check that it is in good order, is not bent, and contacts both plates well enough to prevent damage when the pulley hits it.

Continue with *Steps 36, 37, 38,* and *39.*

In this movement the strip-type verge, common to the ogee and most other American clocks, is replaced by a solid version cut from solid steel. Owing to its shape, it is known as an anchor rather than a verge.

The next step, *Step 42,* is to carefully set up the escapement, including, of course, the anchor. (This is fully explained in Part IIII; see **Escapement** of the "Deadbeat—Solid Pallet—American" type.)

Not all deadbeat escapements will run in the manner described in *Step 43:* When handheld, they generally require the restoring action of the pendulum suspension spring to maintain oscillation.

Continue with *Steps 44* and *45.*

Before putting the movement back in its case, it is important to check the pendulum, for this lies behind the movement and is therefore less accessible than in the ogee. (See **Pendulum** in Part IIII.)

When the pendulum is swinging correctly, without any sign of a wobble or other aberra-

Fig. 9. Seth Thomas Regulator No. 2 8-Day Weight Movement

tion, replace the movement and weight in the case and complete *Step 49.*

Finally, the waiting period in *Step 50* will prove its value, for the adjustments on a deadbeat escapement are somewhat finer than on the average American recoil type.

BANJO 8-DAY WEIGHT MOVEMENT

Case: Fig. 1f   Weight: 7 lbs

There are several sizes of weight banjo clock; because they are somewhat rarely found, they are not covered extensively here. Too many variations in pendulum length and bob weight exist for these data to be given above, but the dimensions of 22½" and 4 oz., respectively, are not uncommon.

Banjo cases vary somewhat in appearance. Some are plain and unadorned, others have veneer inlays or panels of contrasting woods. Some have brass sidepieces, and are quite ornate. Most are T.O., but a few are T & S. The clock is generally of good quality, but the layout is rather congested.

T.O. banjo movements are very similar in design to the Seth Thomas no. 2, save that the crutch wire is brought out in front of the movement instead of behind it. Due to case restrictions, the pendulum must be on the vertical center line, which necessitates a yoke to bypass the hand arbors.

The rectangular movement often rests on the tops of the case sides, and is secured to the case back by a single screw. Because the end-grain of the sides frequently has eroded and the movement become loose, this has proved a poor method of mounting; reinforcing is often required. A good method is to tap the four corners of the back plate—that is, cut a thread in a drilled holed—and then fit brass strips across the top and bottom of the back of the plate (over the holes), using four flatheaded screws in the countersunk holes. These strips should measure about ½" wide × 1/16" thick, and should project about ½" from the edges of the back plate. Four wood screws through these projections will then secure the movement to the case. Where necessary, drill relief holes in the straps so that they do not interfere with any projecting pivots. If this correction does not meet with total approval on grounds of originality, the best that can be done is to strengthen the existing arrangement. The tops of the case sides can be reinforced with brass, and the worn and enlarged single screw-hole redrilled after being closed up with a mixture of epoxy resin and fine sawdust.

Except in a very few details, banjo movement repair is identical to that of the Seth Thomas no. 2 regulator. Some of those details are as follows: The clutch on the minute arbor may be near the rear pivot; the hour pipe may have a slightly different configuration; lantern or cut pinions may be used. Variations such as these amount to maker's preferences rather than differences in design.

The lead weight of a banjo is trapezoidal in shape and has a semicircular pulley slot in the top for greater compactness. Its dimensions are such that it will slide freely up to the top of the case's tapered throat. As the pulley angle changes, compass-fashion, with the rise and fall of the weight, the weight tends to twist. A thin partition is therefore installed in the restricted space between the weight and pendulum to prevent their interaction.

Fig. 10 shows a wooden mould with a wire (coathanger) hook installed, ready for casting a lead banjo weight. The head of the hook lies inside the wooden semicircle and will therefore remain outside the cast lead; the bends in the hook ensure that it will not pull out of the lead. Molten lead will char a wooden mould slightly, but repeated use is nonetheless feasible, as the carbon layer tends to insulate the wood somewhat from subsequent pourings. The mould should be strongly nailed together and free from cracks and holes, thus avoiding undue flashing removal. The wood should be dry to avoid excessive bubbling of the lead. The bottom end of the hook is cut off flush when the casting is cold.

The shape and measurements in Fig. 10 are typical, but naturally they will not suit all clocks. The width of the throat should be measured twice: at its top, and again some 6" down, depending upon the required poundage. The front-to-rear thickness of the throat must also be measured. Reducing each of these three values by 1/8" to 3/16" will give the internal dimensions of the mould, aside from its length.

The volume of lead is simple to calculate. The area of a trapezoid is the product of its height times half the sum of its unequal widths ($h \times \frac{1}{2}[w_1 + w_2]$). Multiplying this figure by the thickness of the weight will give the volume. The volume of the semicircular cutout must be subtracted from this overall figure, the formula being $\frac{1}{2}\pi r^2 \times t$, where t is the thickness of the weight, r the radius of the half cir-

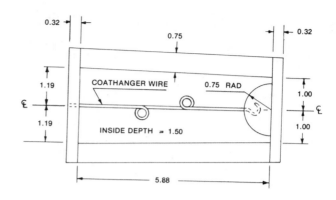

Fig. 10. Banjo Weight Mould

cle, and $\pi \cong 3.14$. The resulting volume in cubic inches is then multiplied by 0.41 (the weight in pounds of a cubic inch of lead), to give the final poundage.

MISSION LONGCASE 8-DAY WEIGHT MOVEMENT

Movement: Fig. 11	Case: Fig. 1g
Crank size: 6 or 7	Weight: 6½ lbs
Pendulum length: 37¾"	Bob weight: 1¼ lbs

Mission clocks are made in mantel, wall and longcase types. Their open, lattice-like cases are generally made of oak slats nailed or screwed together, with mortised corner joints being used in the longcase versions.

Access to the movement is very easy. It is only necessary to take off the hands, and then remove the dial by turning the four convenient turnbuckles screwed to its back. The hands are usually of solid brass, about 0.10" thick. The minute hand, including its very necessary counterweight, weighs about 1½ oz. In anything short of a street clock, this might well be termed heavyhanded.

Aside from the points mentioned below, the basic repair text applies, layout and size being the chief differences between the two types.

The brass-encased weights are usually lead filled, but sometimes a pound or so of sand is also included. The weights are not compounded, but pull directly on the ladder **chains** which pass over sprockets on the no. 1 arbors. These chains should be cleaned along with the rest of the movement, after which they must be carefully examined, link by link, for distortions. It only takes one bent link to cause a jam. In correcting crooked links, you may have to expand their loops, but do not do so to a point where the links become stiff; the links must be expanded wide enough to slip easily over the sprocket teeth, but must still be free and loose in relation to one another.

Straighten any chain that twists when hanging, by applying a countertwist while the chain is under tension. The free ends of the chains must be fastened to wooden bobbins, or similar devices, that are too large to pass through the seatboard holes when the clock is run down. These prevent the chains from running right through the sprockets should the weights fail to reach the floor first. On some of these clocks, the weights are cranked up; on others, the weights are pulled up manually by the chains, which usually have from 38 to 40 links per foot.

The strike hammer on these clocks is fairly long and heavy, and it hardly needs the assistance of the weak spring coiled around the hammer arbor. Should the strike train hesitate to run, then one end of this spring can be unhooked to correct the situation. The sound level will not be noticeably affected thereby. The hammer head, on removal, must be put into dish no. 2, as its immersion in the cleaning machine would destroy the leather insert.

Although the mission longcase strike train uses the same general principle as the ogee, there are a few significant differences, as it is a five-arbor train, as against three in the ogee. Setting up the strike is relatively simple, however.

In the ogee, the hammer-lifting pins are on the S1 arbor and the slotted cam is on the S2 arbor, which means that these two arbors require synchronizing. In the five-arbor-train movement, the hammer-lifting pins, of which there are only two, are on the same arbor (S3) as the slotted cam, so they cannot get out of phase.

In the mission movement the strike train locks when the cam-locking hook drops into the slot in the cam. After reassembly the count hook may have to be adjusted to obtain a correct dropping action (see Fig. 7d). When this has been achieved, apply finger pressure on the S3 gear in the direction in which the train would normally run, and allow the train to lock. In this position, the warning pin on the S4 gear must be at roughly twelve o'clock—i.e., close to the top edge of the rear plate. It must not be anywhere near the warning hook, which moves up from below when actuated by the unlocking hook. Should the pin not be in this position, gently pry the plates apart, as in *Step 41,* and remesh the S4 pinion with the S3 gear. This is easiest to accomplish with the fly removed. There is only one critical meshing in this movement, whereas there are two in the ogee.

The escapement is generally of the deadbeat type. To set it up see part IIII, **Escapement** of the "Deadbeat—Solid Pallet—American" type.

Fig. 11. Mission Longcase 8-Day Weight Movement

Fig. 12. American Mantel-Shelf Clocks

a. Tambour

b. Black

c. Gingerbread

d. Doric

e. Steeple (Sharp Gothic)

f. Column

g. Beehive (Round Gothic)

d.

e.

f.

g.

a.

b.

c.

d.

Fig. 13. American Wall Clocks

    a. Depot

    b. Anglo-American

    c. Saxon

    d. Short Drop School

    e. Long Drop School

    f. Wall Gingerbread

    g. Teardrop

8-DAY SPRING MOVEMENT

Movement: Fig. 14	Cases: Figs. 12 & 13
Crank size: 6 & 7	Bob weights: 2–16 oz

Figs. 12 and 13 show a small selection of the wide variety of mantel and wall clocks which house movements similar to the 8-day spring movement shown in Fig. 14. Some of the wall clocks have T.O. movements, but as far as repair is concerned it serves no useful purpose to consider them as anything but simpler versions of the T & S movement. There is no basic difference between the mantel and wall clock movements, the variations being in the trains, pendulum lengths and mountings. In view of this, the gingerbread clock movement of Fig. 14, together with its repair text, can be considered as being typical of all the clocks shown in Figs. 12 and 13.

The movements of the tambour and black mantel clocks (Figs. 12a & b) are front-plate mounted; that is, the mounting lugs are on the front plate. These movements are removed from the back of the case. The tambour clocks sometimes have a **bim bam** strike, on two or three horizontal gong rods. Care must be taken to avoid bending these rods when withdrawing the movement. Although the rods are seemingly rugged, at the mounting block they taper down to a fraction of their visible diameter and hence become very weak (see **gongs**).

Owing to the wide variety of American clock cases, there are naturally many variations in the way the movements are held in place, but despite this the "works" are almost always simple to remove.

Assuming this has been done, consideration will now be given to the movement of Fig. 14. This movement is from a *kitchen clock,* as the makers called it, though today most people seem to prefer the term *gingerbread.* Two other name changes involve the *sharp gothic* (Fig. 12e), now commonly called a *steeple;* and the *round gothic* (Fig. 12g), now known as a *beehive.*

Remove the hands and pendulum suspension rod and set the movement on the bench. Carry out *Steps 6, 7, 8, 9,* and *10.* Always put removable hammer heads in dish no. 2, to avoid destroying any leather inserts in the cleaning machine.

In the clocks previously considered, all driving forces are neutralized when the weights are removed. However, in a spring-driven movement such is not the case.

Step 11, therefore, applies to spring-driven clocks only, and is to let the springs down. Without this precaution Step 12 would result in a disastrous distribution of arbors around

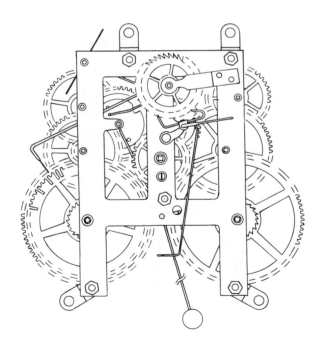

Fig. 14. 8-Day Spring Movement

the work area. ("Dad, how come you're looking for clock parts on top of the curtain rail?")

To be thoroughly cleaned, the mainsprings must go in the cleaning machine in a let-down, or uncoiled, state, for neither the cleaning solution nor the rinse can penetrate between the close turns of a partially wound spring. Always use a **let-down**, or **bench**, **key** to relax the mainsprings. An ordinary clock key is definitely unsafe; its use generally causes damage to both clock and fingers. It is feasible to make a double-ended let-down key (see Fig. 15): Drill a center hole and cut a slot in each end of a 3½"-long, 1"-diameter birch dowel (not softwood), and cement a clock key in place with epoxy resin. The key flanges should then be cut off, and the ends of the dowel secured by a ferrule or lashed with wire and soldered. (See Fig. 15b.)

The job must be strong and sound, for mainsprings command respect; they are not to be taken lightly. Four such keys will suffice—sizes 5, 7, 9, and 11—and the size should be marked on the ends of each handle. Double-ended keys save room in the tool box. They are available from suppliers; Fig. 15a shows a typical configuration. In this instance I prefer the supplier's product, for it is both stronger and neater than a modified clock key. The above method of making let-down keys is given only because an appreciable number of people wish to make their own; and they usually underestimate the forces involved!

To let the mainsprings down, the appropriate bench key, no. 7, is fitted onto the winding square and gripped firmly. It is then turned enough to raise the click, which can then be disengaged from the ratchet gear with a scriber or small screwdriver. Your key hand should then relax its grip enough to allow the spring to unwind in a controlled manner. The last few turns are loosened by pushing the spring sideways out from between the plates, as far as it will go. Some pressure will still remain, but it will not be enough to cause any damage when the plates are separated.

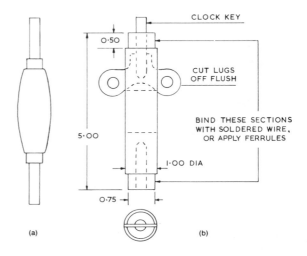

Fig. 15. Let-Down Keys

While this is being done (*Step 12*), keep a spare finger or two on the no. 1 gears to prevent the no. 1 arbors from jumping out of their rear-plate pivot holes. As soon as the front plate is free of the hand arbor and the winding squares, the no. 1 arbors must be removed and the end-loops of the springs lifted off the pillars that hold them. The springs and their arbors go into no. 1 dish, to be cleaned in the machine along with the rest of the movement.

Mainsprings need only be fully relaxed for repair or cleaning. If, for any other reason, clean, freshly greased mainsprings have to be let down (for example, to facilitate strike-train adjustments), use the C-Rings which formerly came on brand-new springs. To reduce costs, manufacturers now often use wire, instead of these very handy C-Rings which were made from round steel rod 0.18" or more in diameter. In the absence of a C-Ring use steel wire of at least 0.050" diameter.

Continue the repair procedure with *Steps 14 through 29*; in Step 16 you will find that the warning pin is on the no. 4 gear instead of the fly arbor. In Step 14 the center gear is not easily removed, so unless the center spring requires attention or damage is evident, the center gear should not be removed from the minute arbor.

Step 30. First scribe a small *S* on the loop of the strike mainspring to identify it. Then examine both springs carefully for cracks, sharp bends, and rust.

A spring that is cracked a few inches from its outer end should simply be shortened, as described under **mainsprings** in Part IIII. However, if the crack is anywhere else a new spring is required. Never leave a doubtful mainspring in a movement; it will inevitably break later, causing damage to arbors and gear teeth. Unduly sharp bends are sometimes present in the center of the spring, and these must be examined closely for incipient cracks. Adjustments with round, wire-bending pliers are sometimes feasible; long-nosed pliers must be used with care, lest their sharp edges aggravate the problem.

Rust is removed with emery cloth, the spring being stretched out from a vise-held, 3″ common nail which passes through the end-loop of the spring. Many of the old, original mainsprings are still in use; they have a pitted, blacksmith-shop sort of appearance, which is not amenable to smoothing. Despite this, they generally continue to work well, and should not be replaced unless obviously weak.

When the spring is clean and dry, it is an easy matter to apply a thin coating of spring grease. Still holding the spring on the nail, dip a fingertip in the vial and run finger and thumb along the spring. The grease should cover the entire length of the spring, but only a thin coating is required. The close center coils can be lubricated by means of a splinter of wood and a few drops of clock oil. Clock oil is sometimes used as a spring lubricant, but on the broad surfaces of an open spring it does not last as long as spring grease.

If the mainspring is somewhat loose on its arbor, the central coil must be closed up a little. Careful use of long-nosed pliers is called for here. Coils too close together to allow entry of plier tips can be separated by judicious use of two scribers. These are placed 180° apart, and forced between the turns. A combination of these two methods will obtain a good shape for the inner coils and a good fit on the arbor.

Should the spring slip at all, enlarge the central coil with a scriber and remove the spring from the arbor. Now examine the spring hook: It must be tight in the arbor, while projecting sufficiently to engage the spring; a small undercut should be evident. If the hook is not undercut, check the ratchet for the winding direction, and, using a half-round Swiss file, undercut the root of the hook on the appropriate side. This will ensure that the spring does not slip off the hook and that increased pressure will only draw it closer to the arbor. Fit the spring back on its arbor—in the correct sense, so that it winds on correctly—and firm up the inner coil as before.

As a long, unwound, greased mainspring occupies considerable space and has a marked affinity for dirt, it is best to partially rewind it without delay. Except in a very few cases, an untamed mainspring makes reassembly needlessly difficult. There is little need of a **mainspring winder** for open mainsprings, for the clock movement itself serves very well as a winder, as shown in Fig. 16.

At this point, check that the mainspring retaining studs are tight and firm in the back plate. They stop run-down springs from pressing on adjacent arbors. Should they be loose,

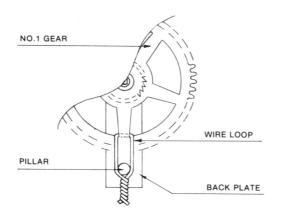

Fig. 16. Winding a Mainspring

tighten them along the lines shown in Fig. 5. If the studs are crooked, bend them gently upright with a hollow punch or pair of smooth pliers.

To complete *Step 30*, fit both T & S no. 1 arbors into their correct positions in the back plate, and set the spring loops over the pillars or posts which originally held them. See that the outer coil of each spring is inside its spring retaining studs, and then fit the top plate on. Secure the plates with nuts, screws, or pins, whichever is applicable. Now fasten one of the gear spokes of each no. 1 gear to a pillar with a short length of sound steel wire not less than 0.050" thick (see Fig. 16). Check that there is no chance of this restraining wire breaking or loosening, and then wind the springs up until a C-Ring can be fitted over the coils.

If C-Rings are not available, pass a wire of not less than 0.050" diameter around the spring and its pillar, and twist the ends securely together.

Let the spring down with a bench key, in the same manner as described earlier. Now cut and remove the spoke-holding wire, leaving the second wire in place to restrain the spring. Do not use the same wire twice, as the twisted section will be much weakened.

The no. 1 arbors, together with their wire-held mainsprings, can easily be removed as one unit if required.

Continue with *Steps 31* and *32*. However, if the escapement lies between the plates, jump ahead to Step 42 (page 48) before going any further. Then return to Step 31.

Step 33, replacing the center gear on its arbor, will not be necessary if it was not removed in the first place (Step 14). Otherwise, follow the ogee text for this step.

In *Steps 33–40*, use Fig. 14 instead of Fig. 5.

The strike train is almost identical to that of the mission longcase 8-day weight movement, and is set up in the same way. The chief variation among 8-day T & S spring movements, aside from relatively minor differences in layout and gear trains, is in the manner of locking

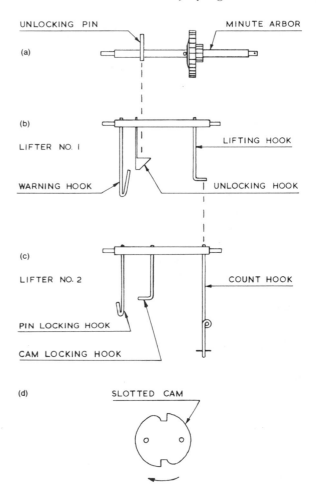

Fig. 17. 8-Day Spring Strike Components

the strike train. In one type of spring movement—let us call it *version no. 1*—the slotted cam actually does lock the train, as it does in the previously described mission longcase movement. In this case the slotted cam comes close to deserving its old title of *locking plate*. However, it is good engineering practice to control clock gear trains at points of minimum force—that is, as high up the train as possible. It is better, therefore, to use the warning pin on the S4 gear as a stop-start device, than the cam on the S3 arbor. Despite the increased speed of the S4 arbor, the force required to lock and unlock the train is very much less. This principle is used in what I shall term *version no. 2*.

Synchronism between cam and count gear is obtained as before, by bending the end of the count hook (Fig. 7d). As in version no. 1, lifter no. 1 (Fig. 17b) raises lifter no. 2 (Fig. 17c), but there are now six hooks instead of five, and they are located differently. Also, in most models of version no. 2 the unlocking hook is designed so that damage no longer occurs when the hands are turned backwards.

In both versions, the cam has two slots, which have either flats or rounded exit corners to enable the train to run when the unlocking hook drops.

The first thing to remember when setting up any spring-driven strike train, is to let down all mainsprings into the care of the C-Rings or wire loops, making certain that there is absolutely no energy left in them whatsoever.

To proceed with *Step 41* on a version no. 2 movement, set the count hook up as for the ogee, but with the cam-locking hook centrally located in the cam slot. In this position the warning pin on the S4 gear must again be at twelve o'clock—i.e., close to the top of its circle. This positioning is slightly more critical than in version no. 1, as the pin must rest against the pin-locking hook to lock the train.

This situation is obtained in the usual way, by removing the fly and adjusting the meshing of the S3 gear and the S4 pinion. The train can be checked manually, with perhaps half a turn of windup on the spring to prevent it from unhooking.

Complete *Step 42* next, again with half a turn on the mainspring. When both trains are set up ready to run, it is time to remove the C-Rings or wire. This is done by simply winding up both mainsprings until the C-Rings can be taken off or the wires cut; never cut a wire under tension, or damage may occur.

Step 45, oiling, is next.

Then comes *Step 47*, reinstalling the movement in its case. If the wood screws holding the movement have worn out their welcome in the back of the case, rotate the lugs and put the screws into fresh wood. (It is good practice to plug the old, worn holes with pine pegs and glue, leveling off both sides afterwards; care should be taken, however, not to ruin any original paper that may be glued to the outside of the case back.) The new holes must maintain the movement in a position compatible with the dial. The dial should be centered in the door opening in cases such as those depicted in Figs. 12c, d, e, f, & g, with the numerals III and IX lying horizontally. In wood-fronted cases such as Figs. 12a & b and Fig. 13b the winding squares and hand arbors must all be centered in their respective dial holes.

Install the strike-setting wire where possible (*Step 48*).

Before proceeding to Step 50, complete Step 42 by giving the crutch wire a final bending adjustment to set the clock in beat; do this on a level surface.

Step 50 completes the repair.

If you consider that further adjustments may be required, you should test the movement for a few days on the bench, instead of putting it in its case. (See **Clock stand** in Part IIII for details.)

30-HOUR WEIGHT WOODEN MOVEMENT

| Movement: Fig. 19 | Case: Fig. 18 |
| Crank size: 6 | Weights: 2½ & 3¼ lbs |

The movement in Fig. 19 is a typical one of several variants, all of which are known as *wooden works*. Most of them are 30-hour T & S movements, and they are most commonly found in cases similar to Fig. 18, the "column and splat," as it is known. More interesting cases, perhaps with carved columns and feet, are found less often, and, like the venerable pillar-and-scroll type, are more highly regarded by collectors. There are many other

Column and Splat

Fig. 18. Wooden Works Clock

Fig. 19. 30-Hour Weight Wooden Movement

case variants, which usually involve changes in the proportions of the pillars and in the shape of the top.

Although other clocks have been dealt with earlier in this book, it must be realized that the wooden works clock appeared first on the scene. Since it is the forerunner of the ogee, it is not surprising that it utilizes the same general idea—i.e., a rectangular case with a weight at each side of the movement. In general, the later, metal movements owe much to their wooden ancestors.

Whatever combination of time, strike, and alarm trains is used, all wooden works utilize similar techniques and components. Gear trains and pendulum lengths vary, and escapements and suspension springs are to be found in several positions. However, the basic concept of wooden plates, arbors, and gears with steel pivots, lifters, and pendulum shafts, is common to all of them. The escapement of these clocks is made of brass and steel, as usual, although wood is used for the escape arbor. Repair techniques differ sufficiently from those used on all-metal movements to warrant a mostly independent text, despite the presence of similar wear problems.

The first few steps of repairing a wooden works are as in the ogee text. Perform *Steps 1–3.*

In *Step 4,* the weights are easily removed at almost any stage of winding, without the need of disengaging the clicks.

Step 5, removing the movement, is generally easy to accomplish. The movement is usually held in place by three or four pins which run from the large holes in the front plate into vertical wooden ribs, more or less as in the ogee. Mounting methods vary somewhat; sometimes the back plate of the movement sits in a cutout in the case back, with a cover of tinned steel over it to protect the pivots. In other cases, the back is solid, and the movement may sit forward a little to give clearance to any projecting pivots, etc. Often these original mountings have become very worn and loose, and they may have been supplemented with nuts and bolts.

The oak plates of the movement do not accumulate oily dirt as do brass plates, and solvents must not be used on them in *Step 6.* It is enough to brush off any dust, using a vacuum cleaner if necessary.

In *Step 7,* check the movement for completeness, wear, and damage.

Five arbors per train is typical for wooden works. The strike train has characteristics that appear also in the ogee and the 8-day T & S movements. Write down the locations of worn pivots, etc., as in *Step 8,* as usual. Pivot hole damage by center punch addicts is not too likely in these movements, as the ivory discs would have been smashed and the brass ones would either have come loose or splintered their sockets.

In *Step 9,* note that the fly is not used as a strike-train control here. The relative positions of the hammer, the cam on the S3 arbor, and the warning pin on the S4 gear must all be noted, should any doubt about reassembly exist. It is always a good plan to repeatedly actuate the strike train of an unfamiliar movement, observing its action attentively until it is thoroughly understood.

There is no need to mark the count gear in *Step 10,* as it will not mesh if it is wrongly assembled. As it is conveniently held on the outside of the front plate by a swinging wire spring, it is easy to remove and remesh.

The two no. 1 arbors are very much alike, so both the S1 drum and S1 gear must be marked with an S, as usual. If there is any doubt about the time-side arbor positions, the gears can easily be penciled T1, T2, T3, etc.

Step 12. Remove the four or five tapered wooden pillar pins and dismantle the movement. Handle all parts with care, for wood does not possess the strength and rigidity of metal.

Take off the count gear, as described immediately above. There is no need for the usual two dishes to store parts, as the cleaning machine will not be used.

As oil is little used in wooden works, the components generally stay fairly clean, and they can be simply brushed off or vacuumed to remove any dust that has gathered. A small brass brush is fine for cleaning up the escape gear; check both sides, and carefully straighten any obviously bent teeth.

Step 16. Continue by inspecting all gears and pinions. Due to the effect of wood grain, broken teeth occur more often in wooden works than in metal movements. Two methods of repairing broken teeth are given in Part IIII under **Tooth replacement: Wooden gear.**

Step 18. Wooden pinions are generally not a source of trouble. Aside from the crutch wire on the verge, there is very little to solder here.

The fly is relatively fragile and can easily be broken along its grain. It must therefore be handled carefully in *Step 19.* Repairs to the fly and other wooden parts are best carried out with epoxy resin rather than wood glue; a much stronger bond will result. Do not attempt to put much pressure on the fly friction spring, or breakage is likely to occur.

Step 20. The hammer-lifting pins are set into the S2 gear and must be inspected for excessive wear. Loose pins are not often to be found; when they are, they must be either made firm with epoxy or replaced with pins of a slightly larger diameter.

Epoxy resin will also serve well to firm up the pendulum cock and the escape cock (*Steps 21* and *22*) in addition to any lifters that may be loose in their arbors.

As the pallet pin is mounted in brass, it is tightened as in *Step 23,* save that the brass disc is first removed from its recess in the front plate. The clock pillars naturally cannot be riveted, but any loose ones should be made secure with glue or epoxy. Do not use nails; there is always a risk of splitting the wood.

Step 24. Rust on wire components such as the lifters can be removed as described previously; however, the hook itself—not the

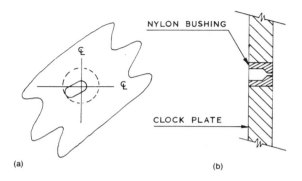

Fig. 20. Nylon Bushings for Wooden Plates

wooden arbor—must be clamped in the vise.

Step 25 is unchanged except that here there is no boss on the hour hand, which facilitates the removal of rust from the hands.

Step 26. The winding clicks on the no. 1 gears are simple to deal with in most cases. If a click is broken or missing, a new one can be readily made up from a small piece of birch. The measurements are approximately ½" × ¼" × ⅛". A centered saw cut is made the length of a long edge, to receive the click spring. A well-fitting hole, for a firmly placed bearing pin, is required. The free end of the click must mesh well with the teeth of the ratchet gear. (See also **Click** in Part IIII.)

Step 27, dealing with the pivots, is relatively easy. The steel pivots are simply pulled out of their holes in the wooden arbors, polished up in the lathe, and reinserted. The pivots must not be extracted with pliers in the usual manner, however, as the resultant scoring would negate all benefits. Use several layers of masking tape on the gripping surfaces of a pair of smooth pliers for both removal and replacement of pivots.

If a pivot is already too badly scored to polish, simply reverse it and polish the other end, taking care to round off the tip. If the worn end does not fit snugly in the arbor, a little glue in the hole should suffice. However, epoxy resin is too permanent for this particular job, as the pivot may have to be taken out again sometime. Bent pivots must be straightened with great care, for they are strong enough to split the arbor, should it be in any way weakened. Remove or tape down the cables on the winding drums, then mount the no. 1 arbors in the lathe and polish their pivots as usual, with a polishing stick.

There are three types of bearing involved in wooden works: a simple hole in the plate, an inset ivory disc, and an inset brass disc. As the plates are nearly 5/16" thick, almost all the pivot holes are recessed to give a shorter bearing. Exceptions to this include holes for the no. 1 arbors and those lifters that have wooden pivots; these bearings usually run the full thickness of the plate.

Step 28 for wooden movements is therefore threefold. A good way of repairing the first type of pivot hole is to use a nylon bushing. A bushing made of a dense hard wood, such as birch, will also serve, but I prefer nylon, especially when loaded with molybdenum disulphide. However, there are no doubt other suitable materials in today's galaxy of synthetics.

Fig. 20a shows a well-tried method of maintaining a reference center when enlarging a worn hole so that it can take a bushing. In many instances, this enlargement will coincide with the recessed hole on the outside of the plate, but it is by no means guaranteed to do so.

Having determined exactly where the pivot hole center should be, take a sharp pencil and rule two fine lines, at 90° to each other, which meet at that center. Were the plates of metal, a scriber would be used instead. It is essential that the eyes be trained so that such a layout can be done quickly and accurately, as measurements are hardly feasible; naturally, such skills come only with much pratice.

Figure. 20a shows a typical instance in which a hole has been worn away from its true center. With the center determined by the crossed lines, file the hole outwards until it almost coincides with the dotted circle. Now round out the hole with the smallest drill that can effectively remove a further small amount of material. The resultant hole, centered on the crossed lines, should be circular and just large enough to encompass the original oval hole.

Turn up the nylon bushing of Fig. 20b on a lathe, and make its O.D. (outside diameter) a press fit into the hole in the plate. On the lathe, make the recess hole and the center hole—pivot hole—of the bushing. The latter must be about 0.005" less in diameter than the pivot for which it is intended. As tapered broaches do not work well in nylon, the hole must be drilled to size in small increments, until it suits the pivot. Sharp drills running at a fairly high speed with a slow feed will result in a smooth enough hole.

In the absence of a lathe, the O.D. of the nylon bushing cannot be altered, and the hole in the plate must be filed out to a press fit for whatever nylon rod is available. Before sawing off a piece of rod, you should drill the bushing's recess hole as centrally as possible in the end of the rod. The rod is then pressed into the hole in the plate, sawed off, and filed flush on both sides. The two centering lines are now carefully extended to give the original center position again—on the nylon bushing. The pivot hole is then drilled out as described above. (Drilling the recess hole at this stage instead of previously would very likely pull the bushing out again, or ruin it by running right through.)

Although satisfactory when carefully done, the latter method is much less precise than a turned job: A lathe soon proves to be essential for working on clocks. In either case, the pivot hole must center on the intersection of the lines, as any drift will adversely affect arbor position and thus gear meshing.

As an ivory bearing is not particularly good mechanically—and is particularly bad from the elephant's point of view—a worn ivory bushing disc is best replaced with one made of brass or nylon. The first thing to do is to mark again the true center of the pivot hole (Fig. 20a). The worn disc is pushed out and measured for diameter and thickness. A new bushing disc is then turned up on the lathe, pushed into place, and the center hole scribed in and drilled, all as described above.

Although brass and nylon discs are functionally interchangeable, it seems consistent to replace ivory with nylon or some other nonmetallic material, and to replace brass with brass. In most cases, a worn pivot hole in a brass disc can be bushed as described under **Bushing** in Part IIII—that is, in the same way as brass movements, the disc being removed for the purpose.

Step 29. When the bushing is complete, it is time to clean the pivot holes. As with Step 28, the methods used vary with the type of bearing.

For the simple hole-in-the-wood type, twirl a tapered smoothing broach a few times in the hole under light pressure. This will make the hole smooth and more circular. A cutting broach will clean the hole out, certainly, but will leave it slightly rougher, so it must be used sparingly if at all. As oil is not used on wood or ivory bearings, there are no black deposits to be removed. Holes that do not require bushing are generally quite smooth, and so require little work.

Ivory discs usually clean up well with pegwood. This must be inserted from the disc side of the plate, as pressure from the other side can push the disc out of its recess. Should pegwood prove inadequate, a few light twirls of a tapered cutting broach will once again be effective, but care must be taken that the hole not be significantly enlarged thereby.

Brass discs are cleaned the same as ivory as far as pivot holes are concerned; however, with brass, there is usually some verdigris present. Because wood is somewhat hygroscopic, whenever it comes into close proximity with brass the stored moisture in the wood corrodes the brass to form verdigris. For the same reason a little rust may be found on the steel pivots.

Removal of verdigris and dirt from both sides of ivory and brass discs is easily accomplished with a sharp-edged screwdriver about 0.2" wide. A broken flat Swiss file, with the end ground flat, is an excellent fit into the recess holes in the plates, and a few twists soon remove all deposits. (This device is easier to handle than a screwdriver; provided that the end is kept square and sharp, it makes a very good scraper, not only for this task, but for removing excess solder from other jobs.)

Step 31. In checking the meshing of the gears and pinions, always pay particular attention to those involved in previous bushing operations. Because wood shrinks and warps to a varying extent, the meshing of wooden gears and pinions cannot be as precise or consistent as that of metal components. It is therefore to be expected that the depth of meshing will change somewhat with rotation. This is not serious unless it results in a bind at one or more specific gear-tooth-to-pinion-leaf contacts. If this happens in several places, the arbors are almost certainly too close together. Should one or more of the four discs involved be slightly eccentric, rotating them may move the arbors and thus change the meshing enough to correct matters. If only one tooth binds repeatedly, then it is probably warped or a trifle too long.

Another form of bind may occur when the arbors are too far apart. This will sometimes cause one or more gear teeth to abut the pinion leaves in a more-or-less head-on configuration. Rather than lying side by side, the tooth and the pinion leaf halt the train by tip-to-tip contact, in several places around the gear. The cure is to bring the arbors closer by changing the discs, or rotating them if eccentricity is present.

When only one tooth exhibits this type of abutment, that tooth is generally too short. The tooth should be renewed if it is much too short. But often a small buildup of epoxy resin is enough to enable the train to run. Even when correctly set up, wooden gearing has a noticeably greater backlash than metal gearing.

The end play in wooden works will be found to be appreciably greater than that of an average metal movement, shifts of the order of 0.050" being quite common. As the pivots are substantial and not subject to any great stress, this large end play does not pose a problem. In fact it is beneficial, as any slight warping of the plates is less likely to jam the trains.

With wooden works, *Step 33* must be combined with Step 44, for it will be difficult to fit new cables later. In Step 33 the center gear and the no. 1 gears must be reassembled on their arbors in the same way as the ogee. Do not set the friction springs up too tightly, as this may cause undue wear and perhaps damage when setting the hands to time.

Step 34 is as with the ogee. However, it is enough for now simply to set the count gear in place, for it may need remeshing when the strike is set up later.

Step 35. See that the hammer spring is not loose in the clock plate, and that the head is tight on its arm.

Continue with *Steps 36* and *37*.

In *Step 38,* the escape arbor is assembled with the rest of the arbors, as a clearance hole is provided in the front plate for the escape gear.

The relative positions of the lifters are easy to determine, but if in doubt refer to Fig. 21 to ascertain the sequence of operations. In Fig. 21, as in Fig. 7: a lifts b, and b lifts c. Bearing this in mind, and remembering that the count hook meshes with the count gear, greatly facilitates positioning the lifters correctly.

The count hook operates through a roughly oval hole in the front plate. As the front plate is lowered into place, it may be helpful to take lifter no. 2 out of its pivot hole and turn it a little to allow the count hook to pass through its hole. As always, the count-hook blade must lie radially in the center of the count-gear slots (Fig. 21d).

Backlash and end play have been dealt with in Step 31; the chief concern was the effect on them caused by bushing. Naturally the same parameters apply throughout the movement, and all other arbors and meshings must now be checked as with the ogee, in *Step 39*. Fig. 6 applies only to lantern pinions, not to the present cut pinions, for which a line-of-sight meshing guide cannot be given. The comments in Step 39 of the ogee text apply equally to metal or wooden gear trains, for the same principles apply to both types.

Step 40. Check that the hammer spring operates satisfactorily. It must not be too powerful,

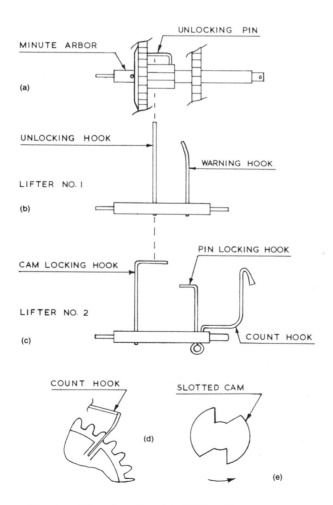

Fig. 21. Wooden Works Strike Components

but it must hold the hammer steady when the train is at rest.

With regard to the strike train, Fig. 21 shows that there are five hooks on the lifters in a typical wooden works. Locking and unlocking is achieved in the same way as in the six-hooked version no. 2 of the 8-day T & S spring movement, described on p. 48. The similarity of operation is not readily apparent, for the two layouts are quite different.

The most obvious difference is in the count wheel, the slots of which do not coincide with the teeth of its integral gear. Aside from the two o'clock segment, which is quite easily broken, this is a neat and reasonably strong arrangement. It is conveniently located on the outside of the front plate, and is easily removed and remeshed, if necessary.

Another difference is that the hammer-lifting pins are on the S2 gear, instead of on the slotted cam. The slotted cam is on the S3 arbor as usual (Fig. 21e). The slots are relatively wide, and have the sloping exit edges typical of non-locking cams.

A more interesting difference is the closeness of the wooden works' warning hook and the pin locking hook; they are only about 10° apart, instead of the more common 180° or more. The effect of this is a very short warning; only a click is heard, as there is no time for gear noise to develop.

Step 41. The strike is set up in the same way as version no. 2 of the 8-day T & S spring movement (see p. 48), taking into account the differences described above. Although here there are no mainsprings to worry about, great care must be taken in adjusting the lifters, for casual bending of, say, the count hook can easily loosen it or even split the wooden arbor.

The only addition needed to the version no. 2 schedule is synchronizing the hammer-lifting pins on the S2 gear with the slotted cam on the S3 arbor. In this respect, the wooden movement is rather like the ogee (see pp. 30–31), where a similar operation exists.

After achieving the above synchronization, and with the cam-locking hook centrally positioned in the cam slot, mesh the S4 pinion with the S3 gear so that the warning pin rests against the pin-locking hook. Both this hook and the warning hook should be positioned at about seven o'clock (looking at the S4 gear, as always, from the front).

Step 42, dealing with the verge and the escapement, is explained in Part IIII under **Escapements** ("Recoil—Strip Pallet—American"). However, the arrangement of the wooden works escapement is somewhat more awkward to adjust than that of metal movements. Therefore a few additional words of explanation are in order.

Due to the vertical position of the escape cock, there is no chance of a tiny adjustment to the escape arbor to put the final touch to the depthing. The only adjustment available is the somewhat coarse rotation of the inset brass disc that carries the pallet arbor (or pallet pin). However, two notches, about 180° apart, in the rim of the disc allow two small nails to prevent its rotation while helping to hold it in place.

It is feasible to set the depthing of the exit pallet without worrying too much about the entry pallet. But if the escapement is a long way out of kilter, it is best to remove the nails to allow the disc to rotate freely.

If only a very small adjustment is needed, it is sometimes possible to use two scribers to persuade the nails to lean slightly sideways in opposite directions, and thus obtain a degree or two of rotation.

As the nails are not always effective for their original purpose, I find it better to use two steel, 0-size, round-headed screws, $5/16''$ long in two new positions around the rim, when there is room for them. These are tightened down after all adjustments are completed, so that their heads clamp the rim of the disc and secure it. One or the other of the nails—whichever is in the better position—can usually be put back to prevent rotation, but this is generally unnecessary. In all instances it is essential that the disc be a firm fit into its recess.

When both trains are correctly set up, proceed with *Step 43*. The two trains should run by the weight of the movement when held in the same manner as the ogee.

Step 44 should already have been done, since, owing to the difficulty of fitting new cables when the movement is completely assembled, it was combined with Step 33, at a point when the no. 1 arbor assemblies were readily accessible.

Step 45 is to oil the movement. No oil is required on bearings of wood or ivory, but opinions differ about the advisability of oiling when a wooden works has brass disc bearings. It is my impression that in the past oil has been used on such bearings, for in many instances it has evidently soaked into the wooden arbors. Examination generally shows that years of running have formed smooth and reasonably good end bearings, and that these have become somewhat impervious to further oil penetration. In addition, a dry steel-and-brass bearing soon scores and binds, particularly when under pressure. Because of these reasons, I feel that the brass discs of wooden works must at this stage be lightly oiled. It is advisable to check all such bearings about once a year, to see that they are not running dry.

Whatever the type of bearing, oil the escape-gear pivot (T5F), the pallets, and the verge pivots, for these are fairly standard in arrangement.

Step 46. Check the two wooden pulleys at the top of the case. They must spin freely and be free of chipped edges which might fray the cables. If small pieces are missing from the edges of a pulley, it can be repaired with epoxy resin. New wood with glue is also feasible, but not quite as strong.

Step 47. Reinstall the movement in its case, and make sure that it cannot wobble.

The strike-setting wire hangs down the left side of the movement from the wire loop shown on lifter no. 2, Fig. 21c. To actuate the strike train, pull it downwards, rather than lifting it as with the ogee. If no strike-setting wire is present, synchronization is achieved as explained in *Step 48,* save that here it is only necessary to back off the minute hand to about 45 minutes, instead of 40. The same precautions apply.

Complete the overhaul with *Steps 49* and *50*.

When running, wooden works clocks periodically give out a few brief clicks and creaks. One type of movement, in fact, is known as a "groaner." However, these noises are not a cause for concern. Having made such timely comments for well over a century, these old-timers are not about to change now.

CHAPTER

4

Canadian Clocks

From the small sampling of clocks shown in Fig. 22, it is easy to see that Canadian clocks copied the designs of America rather than England or France. As neither of the latter countries was at all enthusiastic about exporting clock components, or the technology that produced them, this is not surprising.

Although there are wooden works clocks with Canadian labels, their movements were almost certainly imported; it is unlikely that anything other than the case was made in Canada. This activity occurred around 1835.

Brass movements, however, were manufactured in Ontario by the Canada Clock Co. of Whitby (later of Hamilton) from 1872 to 1885. Some of the personnel reorganized as the Hamilton Clock Co. (1877–1882).

The second and most significant attempt to manufacture clocks in Canada was undertaken by Arthur Pequegnat and his family in Berlin, Ontario in 1903, their previous endeavors having been directed towards the production of bicycles. A few of the early wooden works clocks, with names such as Field of Brockville and Barr of Dundas, are still to be found, but even the clocks made in Whitby and Hamilton are uncommon. This leads most clock people to equate *Canadian* with *Pequegnat*. Regrettably, most of the time they prove correct, as so few clock manufacturers have existed in Canada.

The Arthur Pequegnat Clock Co. prospered, making clocks up to about 1940. A current reproduced-clock catalogue shows that a goodly amount of solid oak went into building strong, heavy cases of considerable durability. About eighty or ninety models or variants were made during the life of the company.

Canadian movements are practically identical to their American counterparts, and are repaired in exactly the same way. However, in some instances I have found the placement of the pivot holes to be a trifle less accurate. Aside from this, these movements are of a similar quality to those of America, and just as rugged and reliable.

A distinguishing feature of Pequegnat movements is that their plates were often, but not always, plated with bright nickel. Examples of this are common, whether the plates be brass or steel. American manufacturers, such

Fig. 22. Canadian Clocks

    a. Pantheon

    b. Ogee

    c. Canadian Time

    d. Regulator No. 1

    e. Brandon

as the Wm. L. Gilbert Co. (with whom Pequegnat was associated), also nickel plated some of their steel movements, but they used a dull finish.

The reason for using steel plates was profit: The manufacturer made a few cents on the brass-steel price differential. However, when the steel plates were not bushed with brass, the clocks wore out very rapidly, and the consumer paid dearly for this very poor engineering practice. It is not surprising that only a few of these shoddy movements survive to afflict today's clock repairers; the pattern of wear has to be seen to be believed.

The Pequegnats did not offend in this respect, for when they used steel plates they followed the lead of those American firms that used brass inserts in their steel plates; these movements have survived, and are quite common.

They are still inferior to brass movements, for the pivots of the lifters and the hammer still run in steel. Some form of plating is required to prevent rusting. One American form of plating, which resembles brass, sometimes comes off when the clock is cleaned.

Although the Pequegnats generally used brass for the plates of their movements, they still often plated them with nickel. There was no good reason for this, except, possibly, uniformity of appearance with their similarly plated steel plates. Both types of movement clean up well, for the quality of the bright nickel plating is good.

The more common Pequegnat clocks do not differ enough from American models to warrant a separate repair text, but the much-sought-after regulator no. 1 has interesting technical quirks which deserve mention.

d.

e.

PEQUEGNAT REGULATOR NO. 1 8-DAY WEIGHT MOVEMENT

Movement: Fig. 23	Case: Fig. 22d
Crank size: 6	Weight: 7 lbs
Pendulum length: 22¼″	Bob weight: 1.9 lbs

Aside from the position of the winding hole in the dial and a few small details, this clock is markedly similar in size and appearance to the Seth Thomas no. 2 regulator; it should be serviced likewise. The case is generally made of oak, but some were made of mahogany.

The movement is held by the same type of solid cast-iron mounting as is used on the Seth Thomas no. 2. However, because of the different positions of the arbors, the mounting on some models leaves something to be desired from a mechanical viewpoint. Other models are pinned to the mounting in a sound, conventional manner.

Despite its similarity to the Seth Thomas no. 2, the Pequegnat no. 1 requires a heavier weight to drive it reliably. The original 7-pound weight is 6¼″ long with a 1⅞″ O.D. When this is missing, it is not easy to duplicate, because brass tubing of this size is not as readily available as, say, 1¾″ O.D. and 2″ O.D. Many owners try to substitute the 6-pound weight of a Seth Thomas no. 2 reproduction, but I find this to be inadequate; the clock usually stops after a while, often when the minute hand is rising. Indeed, the original 7-pound weight is less than ideal. A little experimenting shows that 7½ pounds are required for reliable running. If a 6¼″ length of 2″-O.D. brass tubing of 0.062″ wall thickness is filled with lead, the resulting weight will be close to a satisfactory 7¾ pounds.

The greater power required to drive the Pequegnat no. 1 in comparison with the Seth Thomas no. 2, is due, in my opinion, to three small differences in design: First, the crutch wire is longer, and hence springier, than it would be if it left the underside of the pallet arbor and took a more direct path to the pendulum shaft (as in the Seth Thomas). Second, the minute arbor is not turned down to form the usual integral end-bearing where it passes through the front plate. Instead, a brass collet is driven on to the full-size arbor, resulting in a large, brass-to-brass end-bearing of increased friction.

The third point is undoubtedly more important than the other two. Without counterweights, the distinctive Pequegnat hands are too heavy for a regulator-type clock, for they influence the force applied to the train. This can be verified by noting the difference in pendulum-arc length when the minute hand indi-

Fig. 23. Pequegnat Regulator No. 1 8-Day Weight Movement

cates 20 minutes as against that at 50 minutes. Using a weight of 7¾ pounds, the two arcs measure some 2¼" and 1⅞". In view of these considerations I cannot agree that the Pequegnat no. 1 is the "finest office clock made," as the catalogue puts it.

Although of rare occurrence, it is not unknown for a Pequegnat no. 1 to be fitted with custom-made counterweighted hands. However, lighter hands, such as those of the Seth Thomas no. 2, would be more to the point. The whole idea of a weight-driven, deadbeat escapement regulator is to obtain a small, constant-length arc; hence any changes, even though regular, always tend to detract from the clock's timekeeping ability.

CHAPTER 5

English Clocks

A little background information will be of help in accounting for the considerable differences between American and European clocks. The most significant factor appears to be the widely differing philosophies which in times past activated the creation of clocks on the two sides of the Atlantic. Europe initially led the way, during the several centuries prior to the advent of New World clockmaking. This period saw major horological inventions, and English clockmakers achieved a justifiable reputation for excellence. The results of their long years of painstaking endeavor were clocks built along the lines of a battleship—a very elegant, handmade battleship, to be sure—which, unfortunately, only the elite could afford to buy.

In contrast, in the nineteenth century, America developed mass production methods and applied them to a well-designed, no-frills clock movement which worked as well, or even a little better, than its English forerunner. Installed in a simple case, the cost of one of these clocks amounted to about the price of a row of portholes for the above-mentioned battleship—and one did not have to be in the elite to afford it.

Although this tremendous price differential has fluctuated over the last century, to some extent it still exists. Because England did not mass-produce clocks until after 1900, and then with no great enthusiasm, very few of the older specimens look exactly alike. This makes them very difficult to classify; for example, there are so many variations in design that it is hard to tell just where mantel clocks end and bracket clocks begin.

Since covering all English, Scottish, Irish, and Welsh clocks would require a whole book unto itself, I have restricted my choice to those most commonly found, and which have similar movements.

The best known of all these clocks is undoubtedly the longcase, or grandfather, which despite its size has always been popular. A relatively small number of these have three trains or 8-day count-wheel movements. Also few in number are the musical ones, those of 30-day duration, and the early steel birdcage movement, which was superseded by the now almost universal brass-plate-and-pillars arrangement.

Although in the heyday of English clockmaking almost every little hamlet had one or more clockmakers, the similarity of many longcase movements points to the likelihood of common sources of unfinished movements. Old records indicate that the most commonly made movement was the 30-hour count-wheel type, which utilized Huygen's single-weight system. When the 8-day movement became more affordable, these once-numerous movements were downgraded and cast aside. Considerable numbers of 30-hour movements, which happened to be in good cases, have been replaced by 8-day movements from poor or ruined cases. This regrettable practice has accelerated in the last decade, and has resulted in the 8-day clock's becoming the more common of the two. As dials go with their movements and are generally very difficult to exchange, some very attractive dials have been discarded along with their original 30-hour movements.

The four-arbor strike train of the 30-hour movement works along the same lines as that of the ogee. Both have thirteen hammer-lifting pins on the S1 gear, for example; but the layout of the movement is quite different, and the warning pin is on the S3 gear instead of the fly arbor. As there is also a locking pin on the S2 gear, all three strike gears (S1, S2, and S3) must be synchronized.

The three-arbor time train is straightforward. The most interesting feature, which is often also the most troublesome, is the single-weight driving system for both trains. This is fully dealt with under **Huygens system** in Part IIII. (It is named after its inventor.)

Because the power of a mainspring decreases as it runs down, a compensating device, using a chain and a spiral groove of varying diameter, came into being. It is often credited to Jacob Sech of Prague in 1525 or thereabouts, but it seems to have been in existence at least fifty years earlier. Known as a *fusee,* it levels out the extremes of mainspring power until the force applied to the train is roughly the same whether the mainspring is wound up or run down. This power equalizer thus allowed a more constant pressure to be applied to the escapement. In consequence it came to be very deeply entrenched in English clockmaking. Although a costly and troublesome device to make, it was widely used, even spreading to the striking and chiming trains, that they might run more evenly. Used in conjunction with old original mainsprings, a fair degree of compensation was achieved; however, when a modern spring is substituted, overcompensation often results.

The 8-day fusee movement described in this chapter (Fig. 25) weighs 3½ lbs, as compared to 1 lb for a T.O. American gallery clock movement of comparable performance. I would estimate that the fusee movement would take at least ten times as long to make as would the American movement. Despite this, a vast number of English dial clocks made during World War II still had fusee movements of the old, traditional design.

Many of these cost-is-no-object clocks were scrapped after the war, but those that remain, a bonus to clock collectors, can be remade into very acceptable skeleton clocks. This is not a step to be undertaken lightly, however, for much toil is required to design and cut out pleasing plates and machine suitable pillars, and so on. Nevertheless, the essentials are undoubtedly there.

Bracket clocks, often of a high quality, command a high price. Their movements vary so widely that, aside from national characteristics, perhaps only two notable features might be said to be common to some of them. These are the fusee and a striking mechanism similar to that in the great majority of English 8-day longcase movements. (Both of these features are dealt with in the following texts.) It would serve little purpose to attempt a survey of the many variations found in other bracket clocks, none of which is exactly common.

English wall clocks likewise exhibit variety, the odd one even having a movement partially of wood. Many house the same heavy fusee movement as the dial clock, while the larger ones, such as the "Act of Parliament" type, have movements akin to the longcase.

It is interesting to note that wherever possible, European clockmakers fasten their clock dials to the movements, whereas most American dials are secured to the case front. Possibly the reason is climatic, for the hand arbors of

Fig. 24. English Clocks

a. Mantel

b. Fusee Dial

c. Longcase

American clocks imported into England have been known to bind on their dials, due to warping in the appreciably damper conditions, the movements being fastened to the back of the case.

Up to about 1670, all the types of English clock mentioned above, including the even earlier lantern clocks, used what is known as the verge escapement. This escapement was in use as early as 1385, giving both it and its successors, the "anchor"-type escapement, a life span of about three centuries. (The older arrangement is described under **Verge escapement** in Part IIII; see also its governor, the foliot.)

When the superior recoil escapement came along, in about 1670, many movements were modified in its favor. Such modifications are sometimes all too evident, for the job was not always done well. Owing to its shape, the pallet assembly of the recoil escapement was known as an *anchor,* hence the above-mentioned term *anchor escapement.* George Graham's deadbeat escapement, invented in about 1730, is also anchor shaped. For the sake of clarity, however, the term *anchor* is best used only with reference to the shape of a pallet assembly, rather than to a type of escapement. The terms *recoil* and *deadbeat* are accurate and specific and refer to escapements.

Note that the American term *verge* corresponds to the English *anchor* in that both refer to pallet assemblies, even though their shapes differ. It would therefore seem that our old friend *verge,* having spent at least six centuries of a long life in England, will continue to live vigorously in America. It is essential that all such terms be explained and understood, for by no means do they mean the same thing to all people.

Fig. 24a shows a typical factory-made twentieth-century mantel clock. The movement of such a clock closely resembles that of German clocks dating from about 1880. In view of this, a separate repair text is unnecessary, and the information is given in Chapter 7. The only obvious difference between the English and German movements is that the plates of the English version were often left solid—especially the back plate—whereas the German ones generally had cutouts, to save brass. Both cut and lantern pinions are to be found in each of these movements; the cut pinion is more evident in the English movement.

The cases of these clocks (Fig. 24a) are of simple design. Walnut veneer was widely used, sometimes with small amounts of other woods as contrast. The dials are often just a chapter ring pinned to the front of the case and protected by a chromium-plated, brass or steel bezel with a convex glass. Although not in great demand as yet, they may well become collectible in the future.

c.

8-DAY FUSEE MOVEMENT

| Movement: Fig. 25 | Case: Fig. 24b |
| Pendulum length: 7″ | Bob weight: 12 oz. |

When you lift a fusee clock, its unexpected weight makes quite an impression, for everything about it is massive and heavy. Even the bezel is turned up on the lathe from solid brass. To support a movement weighing several pounds, the steel dial undoubtedly needs its 0.050″ to 0.060″ thickness. Even the single fusee, T O dial clock of Fig. 24b weighs about 11 lbs, and the two-fusee T & S versions are heavier still. The latter two-train fusee movements are found in drop dial cases, and have longer pendulums.

There are generally two trap doors in fusee clock cases, one at the bottom for regulation and one at the side behind the dial for removing the one-piece pendulum. The pendulum is mounted on the back of the movement, and it is not always easy to remove, especially if the **chops** of the suspension spring fit too tightly in the cock. In the few instances where this occurs, take the clock gently off the wall, and, leaning it backwards, lay it on the bench on its back; then ease the chops out carefully.

Step 1 is to remove the hands. Use Fig. 2 as a guide, as usual, for the minute hand. A small screw is often used to hold the hour hand onto the very substantial hour pipe, a method which I consider somewhat clumsy and not altogether effective. This single screw will not stop the hour hand from sagging or being pushed down onto the dial, with consequent scoring. If, however, the screw slot is at the tail end of the hour hand, then the hand can rise enough to foul the minute hand. If screws must be used, then at least two, and preferably three, are required. It is obviously better to press fit a split collet on a slimmer hour pipe, as in American and German clocks.

However, the single-screw arrangement is often found in English wall and longcase movements. Unless additional screws are added, the best that can be done is to ensure that the hour hand is a firm fit on the pipe. A loose hole in the hour hand can be closed up by the light use of a small ball hammer, with care taken to keep the hole circular and the encroachment even. After a good fit has been obtained, the hand disc must be made flat and true by means of an anvil and a suitable, large flat punch or drift. Apart from these procedures, the hands are dealt with as in the ogee.

Step 2. Remove the dial by taking out the three or four screws around its edge; both movement and dial come out as a unit. The entire front of the clock is easily removed by pulling out the four wooden pegs that secure it to the case, but the dial may still have to be unscrewed to give access to the movement-retaining pins.

Step 6 is next, when necessary, but keep the solvent out of the mainspring barrel when cleaning the chain.

Steps 7 and *8* are somewhat simpler here, as so few arbors are involved. Note should be made of the run of the chain from the barrel to the fusee. Should no chain or cable be present, examine the fusee groove: If it is square in cross section, a chain is required; if U-shaped, a cable. Other indications may be the existence of cable holes in the barrel and fusee.

Step 11, as with any spring movement, is imperative. (Refer to p. 44.) The mainspring involved is one of the most powerful in the business, being at least twice the strength of the usual American ¾″ spring. Since the click is inaccessible, set a finger on the escape gear and then remove the pendulum cock. Now remove the pallet arbor. Then allow the train to run down until all of the chain is off the fusee.

There remains about one turn of set-up power in the spring (otherwise the clock would not run right down). This force must be released next. Slacken the set-up click screw (Fig. 25, upper left) just enough to allow it to be moved by finger pressure, and let the last turn out of the mainspring with a bench key. Owing to the large size of the square end on the barrel arbor, it may be necessary to make an extra-large bench key for this purpose. But, because only one turn or even less is involved, a no. 16 longcase crank is a feasible alternative.

Remove the reduction-gear cock, and check that its two steel dowels are firm. Unscrew and take out the stopwork block, and check that its finger and spring are in good order.

Now take off the hour pipe, the reduction gear, the minute pipe, and finally the curved center spring which lies behind it on the minute arbor.

Step 12. Push out the pin in the barrel-arbor square, and then take off the ratchet gear and its click. Pull out the pillar-retaining pins, and gently pry the plates apart. Lift out the five arbors, and unhook the chain, first from the fusee, then from the barrel. Note which hook is used where; they are not identical.

Step 13. Attention must now be given to the mainspring. A mainspring in a barrel is probably the most neglected item in clock repairing, possibly because of an inherent human dislike of something that might leap out and inflict grievous bodily harm, as the law aptly puts it. However, as the need for extracting and reinstalling the mainspring is essential and too widespread—indeed unavoidable—to be lightly dismissed, it is dealt with in full under **Mainsprings in Barrels** in Part IIII. (As the bearings of the barrel are also involved, see also **Bushings—Barrels** in Part IIII.

Step 15. When the mainspring is out of the barrel it must go through the cleaning machine, along with all the other clock parts, including the barrel, its cover, and the barrel arbor.

Step 16. Check the gears for damaged or

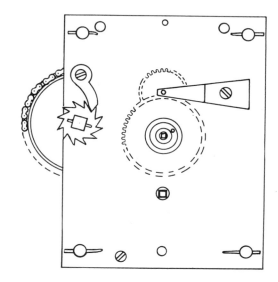

Fig. 25. 8-Day Fusee Movement

broken teeth. No lantern pinions exist here, but the cut pinions must be checked carefully; remove any dirt that has defied the cleaning machine and is still between the leaves of the pinions. Badly notched cut pinions cannot be repaired in the same easy way as lantern pinions—that is, by renewing their trundles. When possible, a good approach with cut pinions is to move the gear that turns them along its arbor a little, so that it runs on a new section of pinion. A new pinion is seldom required.

Step 17. Examine the fusee chain, link by link, to see that there are no bent links or weak rivets. If any of the joints are stiff, mount a 1″-O.D. mild-steel rod vertically in the vise and seesaw the chain round it a few times; in stubborn cases a few drops of penetrating oil will often help, but be sure to clean the chain again afterwards. When all the joints are loose, oil the chain lightly with clock oil.

If a gut line or bronze cable is used instead of a chain, test its strength manually. Discard it if it is at all frayed, and fit a new one later, in Step 26.

Step 21. Fit the pendulum suspension spring into the slot in the pendulum cock, and adjust the two jaws until a firm fit results. The jaws must be free enough to allow the pendulum to pull the suspension spring into line, but not loose enough to allow any wobbling.

Check the pillars for tightness, as explained in *Step 23* of the ogee text.

Treat the hands as described in *Step 25* of the ogee text.

Step 26. The actual operation of the click in a fusee movement cannot be checked visually, as the click is out of sight in a recess in the fusee gear. Check that it is in good order, and clean out any solvents that may have seeped in around it. Take out the small screw that holds the keyholed friction collet in place at the back of the fusee, and push the collet along until the large end of the keyhole allows it to slide off the arbor. Slide the gear off in the same way. Rinse out both pieces.

Pry off the ratchet gear with its three dowels, and clean behind it. Make a note of any small punch marks on the gear and fusee body, to facilitate correct reassembly. Check that the ratchet gear teeth are in good shape, and that the tip of the click is sharp and fits well into them. Test the pin on which the click swivels and make sure that it is firm, strong, and a good fit for the click. Make sure that the click spring exerts a reasonable, but not excessive, pressure on the click. Check the fusee cable; if worn, it must be renewed at this point.

When all is clean, dry, and in order, oil the click pivot, the flat rim of the fusee where it runs on the gear, and underneath the ratchet gear. Oil lasts a long time in such places. Reassemble the fusee and its gear, and put another oil drop on the friction collet where it rotates around the arbor.

Steps 27, 28, 29, and *31,* dealing respectively with pivots, bushings (where necessary), pegging out, and checking the meshing are all done as in the ogee text.

Step 32. A check on the end play of the arbors will show quite a tight clock, with shifts on the order of 0.010"–0.015". Also check for warps in the plates; however, warps are infrequent in these well-made, sturdy movements, and adjustments are seldom required.

Step 36. Clean out the hour pipe, and see that the hour gear is firmly swaged to it.

Step 37. Give the parts a final check before putting the movement together again.

Step 38. Reassembly is relatively simple. Begin by inserting the correct chain hook in the correct direction into its slot in the barrel, and wind the chain around it. It is a good idea to wind all of the chain around the barrel and secure the free end with a loop of thin wire or an elastic band, so that it does not loosen up and get in the way of other arbors.

When the arbors are all in position in the back plate, the hook-post at the end of the fusee groove should be in a position where the chain can be conveniently hooked onto it. This is not, however, a critical matter, for the gears are easily rotated to suit. Next, fit the front plate on, and see that all arbors have end play. Pin the set-up ratchet gear on the barrel arbor, and screw its click loosely in position beside it.

Hook the chain onto the fusee, and rotate the barrel by hand until no more chain will wind on. Holding the chain firmly, rotate the ratchet gear by hand to allow the mainspring to take over the tension, and set the click into the teeth to hold it. Under this light pressure, adjust the turns of the chain as they lie on the barrel to be somewhat closer together at the barrel end opposite the large end of the fusee: This will allow an easier wind-on, which will minimize lateral stress on the chain. The chain will find its own optimum position in the course of normal running.

Step 39. See that all arbors, including those with new bushings, have adequate end play, and that backlash is present throughout the train at all times.

Step 42. Set up the escapement as per the section in Part IIII entitled **Escapements** ("Recoil—Solid Pallet—English"). As there are no built-in adjustments on the average

English fusee movement, it is not an easy escapement to adjust. (The same can be said of English longcase movements, as we shall see, except that the anchor is larger and therefore somewhat more amenable to bending.) A small vertical shift can be obtained for the pallet arbor of either movement by elongating the two holes in the pendulum cock. It is generally necessary to file these holes upwards, so that the cock can be lowered to set up the escapement after the pallets have been stoned, polished, and adjusted.

Owing to its single-screw mounting, there is only one position for the hour hand on its arbor. In contrast, the minute hand has four alternative positions; one of these four must allow the minute hand to point to twelve o'clock when the hour hand is correctly indicating one of the hours. This is easily achieved by the following procedure: Pull the minute pipe forward until the cannon pinion disengages from the reduction gear. The hour hand can then be set to three, six, nine, or twelve o'clock (which can be gauged without the dial), the minute hand to twelve o'clock, and the gear pushed back into mesh. The hand collet and pin must then be fitted to prevent slippage.

Wind the fusee not more than one-quarter of a turn with the usual clock key. Now crank the barrel arbor about three-quarters of a turn, setting the click in with a finger. The train should now run with the movement standing on the bench, without its pendulum. Although the actual amount of setup varies a little according to the lengths of the mainspring and the fusee spiral, it is generally not advisable to exceed one-and-a-quarter turns. If too much setup is given, the mainspring may be fully wound before all of the chain is wound onto the fusee, which would negate the stopwork and endanger the mainspring. Tighten the click screw firmly. Never under any circumstances attempt to alter the amount of setup unless the mainspring is completely run down. A fully wound fusee mainspring is always able

and willing to teach us the respect it richly deserves.

Step 45. Oiling comes next, and is done as usual (see **Oiling** in Part III). Do not forget to oil the barrel bearings. Do not oil the barrel arbor pivots, which remain stationary in the plates. Another oiling point, which is sometimes missed, is the rear pivot of the reduction gear: This is easy to oil from the back of the front plate.

Owing to the method of mounting, the hands will not rotate unless the center spring, which is behind the cannon pinion, is put under tension. After fitting the hour hand, this tension is applied by pushing the minute hand and its arbor firmly towards the front plate and holding it there, while inserting the tapered pin through the hole in the end of the minute arbor. If the pin can be inserted without any such pushing, the minute hand will be too loose and its rotation may be intermittent, thus causing the clock to lose time at a false and irregular rate.

Hand tension can be increased by adding another washer, or by dishing the existing hand collet more deeply. The method of doing this is explained under **Hand collet** in Part IIII.

If adequate hand pressure cannot be obtained, remove all three gears from the front plate and check the center spring. If it is sound, carefully bend the ends up with a pair of wire-bending pliers—a sharp bend will fracture. If it is cracked, make a new one from 0.015″ beryllium copper or phosphor bronze. (Beryllium copper is preferable to hammer-hardened brass, which tends to crack.) This new spring must be diamond shaped, rather than rectangular, to maintain the strength of its center section. Round or square, the center hole must be a close, firm fit on the minute arbor. If square, its sides must be filed parallel to the diagonals of the diamond as this orientation does much to keep the center section as strong as possible. The ends must be curved up—although, as mentioned above, the suggested metals will not take a sharp bend. The center section should be relatively flat.

When hand pressure is satisfactory, fit the dial on and synchronize the hands as before.

Wind the movement up fully, and check that the arm of the stopwork engages the tongue of the fusee, thus preventing overwinding.

Step 47. Fit the movement in its case and hang it on the wall. A good way of levelling a circular clock is to hold a spirit level across the axis going from nine o'clock to three o'clock, and then carefully adjust the crutch arm to suit. The case's four tapered wooden plugs come in handy at this point, for they provide easy access.

If the movement is to go into a **clock stand** for a trial run, it may be necessary to screw the stand down; otherwise the heavy pendulum used in these clocks may cause it to sway, thus stopping the clock.

LONGCASE 8-DAY WEIGHT MOVEMENT

Movement: Fig. 26	Case: Fig. 24c
Crank size: 12–16	Weight: 12–14 lbs
Pendulum length: 40″	Bob weight: 2.5 lbs

Longcase or grandfather clocks evolved from the sixteenth-century European pillar clock via Ahasuerus Fromanteel's 1658 prototype to the familiar shape we know today. Variations in case style, and the woods used in them, are legion, but all use the same general configuration and the common principle of a movement driven by weights which fall inside a hollow column of supporting wood.

The early straight-up-and-down longcase clocks used the verge escapement, with its short bob pendulum. The recoil escapement of 1670, however, allowed the use of the much longer, one-second or "royal" pendulum, so the case had to be widened accordingly. The resulting clock not only kept much better time, but its proportions were, and still are, more pleasing and practical.

Access to the usually heavily built movement in the earliest clocks was by means of lifting the hood upwards, but this was soon abandoned in favor of the almost universal forward-sliding hood.

Generally, the base section of the case is roughly as wide as the hood. The two sides of the narrower waist section extend up into the hood to support the seatboard on which the movement rests.

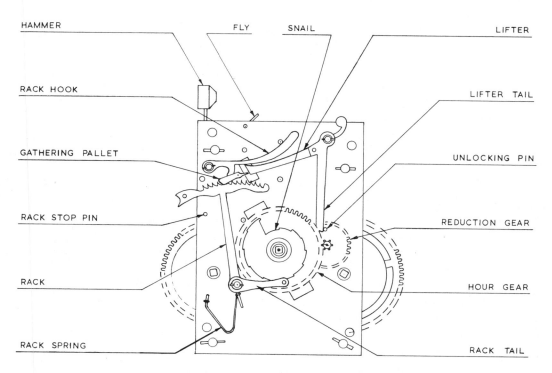

Fig. 26. Longcase 8-Day Weight Movement

After the hood is slid off, the weights must be unhooked from their pulleys and the pendulum removed. The latter is done by lifting the pendulum slightly and then sliding the suspension spring backwards out of the slot in the pendulum cock. In most cases, the movement can now be lifted out; however, some seatboards are nailed down and must be pried up. This usually brings the old square-section nails up with the seatboard; they can be removed after the movement has been taken off. These nails are not really necessary, though they may serve as locating studs for reassembly.

Step 1. Set the movement, dial, and seatboard on the bench in a sloping position, and remove the minute hand and its collet by extracting the pin passing through the tip of the minute arbor. The hour hand is next removed, according to its method of mounting. It may simply be a friction fit, or a small screw may be used. There are two other methods: a U-shaped spring washer which fits into a groove in the hour arbor, or two small pins which run through the corners of a square-ended hour arbor. Set a finger on the escape gear, and take off the second hand with a gentle rotary pull. If there is a calendar hand, it will come off with the dial. (Most of these have a ratchet gear of sixty-two teeth, and are actuated twice a day by a pin on the hour arbor.)

Step 2. The dial is removed by extracting the pins at the rear of the front plate; these may be three or four in number, according to the number of pillars used. In many clocks with white dials, the pillars are pinned to a cast-iron false plate, which has a further set of pillars pinning it to the movement. In this case it is necessary to remove all of the seven or eight pins involved, so that the false plate and the back of the dial can be cleaned. In cleaning, check first for any scribed or pencilled names and repair dates before rubbing too hard.

To digress for a moment, the reason for this clumsy and apparently needless plate lies in the way the clocks were made. Until about 1770, longcase dials were made of engraved brass and, as the chapter ring hid most of the brass-to-brass riveting, no false plate was needed. When the white-painted steel dial came along, the artist took over from the engraver and dials were no longer always made to suit individual movements. The supporting dial pillars had to be riveted to the dial before it was painted, so that the paint would hide the riveting. Because the dials were produced in numbers at places remote from the clockmaker who would put the clock together, there was no guarantee that the dial pillars would fit his front plate. (Although the majority of longcase movements are similar in layout, the actual dimensions vary considerably.) For this reason, a false plate—a sort of adapter plate—was sold, undrilled, to the clockmaker, who then drilled holes in it to suit the dial he had bought. He then fitted further pillars on the back of it to suit the front plate of his movement. The false plate thus allowed almost any painted dial to be fitted to almost any movement.

The maker of the false plate often cast his name into it: "S. Baker—Birmingham," for example. When the front of the dial is unsigned, it is frequently and wrongly assumed that S. Baker made the whole clock. In actual fact, he only made the false plate. If, however, his name also appears on the back of the dial, he supplied, as dial-maker, both together as a unit.

Even the man whose name appears on the face of the dial by no means always made the whole clock from scratch. Often the movements were factory made, although a good deal of hand fitting seems to have been done and there is no suggestion of mass production along American lines. (The plates of handmade movements usually show traces of scribed layout lines and occasionally the odd misplaced hole.)

To return to our procedures: Next, remove the seatboard and dust it off. Sometimes the movement is hook-bolted to it, but more often a tapped hole in the center of the bottom pillars is used—undoubtedly much the better method. It is interesting to note that a few drops of wax are often found sticking to seatboards. Long ago, in some dark corner, someone set a candle there to see what was amiss with the clock.

Step 5. If the cables are worth preserving, untie the knots in them and pull them through their holes in the seatboard, to isolate the movement. In most cases the old gut is frayed, weak, and useless, and should be cut away and discarded.

Continue with *Step 6* if a primary rinse is indicated, and then proceed with *Steps 7 and 8.*

Take out the two screws that hold the pendulum cock, and remove both it and the anchor. This is done by turning the anchor upside down and feeding the crutch wire through the large hole provided for the purpose.

If the crutch wire is broken, it must be silver soldered; in this instance soft soldering is inadequate. (See **Soldering—Silver** in Part IIII.)

It should perhaps be mentioned here that although this heavily constructed, durable movement is not unduly complicated, the time needed for its overhaul is about four or five times that for an American T & S movement. This is because the components are less convenient to repair, and, being appreciably older, are more likely to exhibit greater wear.

Step 9. Make a note of the positions of the strike-train components, using Fig. 26 as a reference. This is always a useful practice, particularly when an unfamiliar movement is involved. It should be realized that there is no absolute guarantee that the strike train is set up correctly, in the first place. As the set-up procedure will be given in detail later, a general idea of its operation is all that is required at present.

Very few 8-day longcase movements use a count wheel; almost all use a rack & snail mechanism. This clever device is generally credited to a churchman named Edward Barlow, although, as with a great many inventions, there is some argument about this. It dates from about 1675; although superior to the count wheel arrangement, it did not entirely supplant it, and both systems continued to be used.

Step 10. There is little need to mark any of the gears, for no two are alike. It is best to scribe the usual "S" on the strike drum and its friction collet, however, for they are hand fitted. The S1 gear might also be marked if doubt exists, but it is smaller than the T1 gear and the two are not interchangeable.

Unpin and remove the rack hook, the lifter, and the rack from the front plate. Also unpin the reduction pinion and, if present, the calendar gear. Slide off the snail and hour-gear assembly together with the rack; these parts may interlock somewhat. If a separate 2:1 calendar gear is present it will also come off, but this gear is frequently missing anyway. (See **Calendars** in Part IIII.) Remove the reduction gear, and take off the bridge by unscrewing its two holding screws, making a note of the sense in which it lies. The bridge is not reversible, and correct reassembly is often ensured by means of two small dowels; a couple of center-punch marks, one on the plate and the other on the bridge, are sometimes used instead.

Slide off the minute pipe with its riveted cannon pinion and then the curved center spring, which lies behind the pinion. Now ease off the **gathering pallet** using two screwdrivers, one on each side of the projecting squared section of the S3 arbor. The two screwdrivers must be similar in size; they are applied parallel to each other, from the four o'clock position, so that their edges engage the underside of the gathering pallet. A gentle twist to both screwdrivers simultaneously will remove the pallet without bending the tapered square on which it is press fitted. The lower screwdriver will rotate clockwise, and the upper one counterclockwise. It is a good idea to use a 1" × ¾" steel shim—say, from the lid of a fruit can—to protect the front plate from scoring. This shim should have a narrow, lengthwise slot put in it about ½" long. It is slid behind the pallet, with the arbor going into the lengthwise slot. A hand-removing tool may be used to do the job, but there is still a danger of bending the arbor with it.

The rack spring may be riveted to the front plate, in which case it must be left alone; if it is screwed on, however, remove it for ease of cleaning.

Step 12. Using the method shown in Fig. 3, take out the pillar pins, gently pry the plates apart, and take out the various arbors. Unscrew and remove the hammer spring. There is no need here to use two dishes for storing parts, unless the hammer happens to have a leather face, which is unlikely.

Step 15. Put all the components through the cleaning process.

Step 16. After cleaning is completed, give all parts a careful inspection. Check the teeth of the gears as for the ogee, and see also that the teeth of the rack are even and of equal height; a short tooth must be replaced or it will cause the strike to malfunction.

Step 18. Check that all collets and gears are firm; soft solder is often employed on the arbors, and play seldom occurs. Make sure that the rack tail is firm on the rack pipe; there must be no play between the rack and the rack tail. Likewise, the lifter tail must be firmly riveted to the lifter. Inspect both parts for cracks.

Step 19. See that the fly is not cracked, and that its bearings are not worn thin. As the brass fly is often frail, any attempt at riveting can result in more cracks. Slim strips of 0.015" or 0.020" brass, soft soldered in place, are generally effective and preserve as much of the original as possible. If a new fly is needed, see **Fly—English longcase** in Part IIII.

Step 20. Check that the following pins are strong and firm: the **rack pin,** the **rack stop pin,** the unlocking pin, the warning pin on the S4 gear, the hammer-lifting pins on the S2 gear, and the short, bevelled pin on the rack tail. On each drum, close to the cable hole there is sometimes an angled pin, which we shall call the cable alignment pin. See that these are firm, and that they slope away from the center of the drum. Their purpose is to ensure that the cable commences to feed onto the drum in a single layer. (However, this largely depends on the tilt of the clock.)

Step 21. Examine the pendulum cock: See that it is not cracked, and that the suspension spring slit is both vertical and, when screwed in position, is at 90° to the back plate. Adjust the jaws of the cock to suit the pendulum suspension spring, as before.

Step 23. Unless the clock has been knocked down or otherwise has suffered severe damage, the pillars usually remain firm in the back plate, but they should always be checked both for firmness and angle. If any pillars are bent, the whole movement will be warped. As considerable force is required to correct this situation, everything but pillars and plates must be removed.

It is sometimes possible to straighten each pillar in turn by means of a bar, two small blocks, and a clamp, until they all line up truly again. This method is time consuming, and as the bends are usually near the plates, it may result in crooked pillars. Those who prefer to save time and live dangerously can simply pin the two plates securely together, rest the edge of one plate on a hardwood block, tilting the whole assemblage slightly, and hammer the furthest (uppermost) edge of the other plate, using another wooden block to protect it. The movement should be held at about 30° from the vertical for this seemingly drastic operation. Oddly enough, the riveting generally remains secure. It is one of the few clockmaking procedures that calls for a 1-lb hammer.

Whichever method is used, when the plates are once again square with each other the pillars must be checked and hammered tight, where necessary.

Step 24. If considerable pitting and rust are evident on the steel components, remove as much as possible with emery cloth. A strip of this can be used for the arbors, hammer arm, etc., and a polishing stick or emery block for flat parts such as the rack, rack hook, and lifter. However, if a smooth, brown patina is present, it is just as well to leave it alone: It is not a good plan to use a rust preventive liquid, as residues will corrode adjacent brass components.

Step 25. Longcase clock hands are rarely made of anything other than steel or brass. Rusty steel hands are dealt with as for the ogee, and brass hands are easily cleaned with very fine emery cloth and a sanding block, followed by polishing.

The minute hand is often found to be bent, due to someone's efforts to make the clock strike on the hour—this is poor practice. Sight along the hand and adjust it until it is dead straight. It is important that no sharp kink occurs at the bottom, which would push the mounting hole off the center line. Lay the hands flat, and sight along them to detect any twists—these are often overlooked, and may cause the hands to jam and so stop the clock. (See **Bluing,** for steel hands, and **Hands—Repair,** for broken hands, in Part IIII.) It is a good idea to lacquer both steel and brass hands after polishing them, as this inhibits rusting and tarnishing.

Step 26. The no. 1 gears are usually somewhat loose, and no longer contact their drums firmly. This is caused by wear on the rigid, brass friction collets, which lack springiness. The collet is sometimes pinned to the arbor, which is a satisfactory method, but in many cases it has instead a large keyhole which enables it to run in a turned slot in the arbor. In this arrangement only a partial bearing is achieved and wear is more rapid.

The keyholed collet is held in place by a small pin; this is often difficult to remove because not enough of it projects to afford a purchase. It is necessary to remove it, however; if other methods fail, it may have to be drilled out. It is not a good way of keeping the collet in place, and is easily improved upon, as will be seen on reassembly.

When the gear has been removed, insert the tips of a pair of small, long-nosed pliers into the large hole in the side of the drum, and extract the remains of the old cable; it is usually made of gut. Several other knots from previous cables are generally still inside the drum, and these too should be teased out: The hole must be held downwards to bring these knots to the hole, where a small wire hook and the pliers can be used to extract them. If they are not removed, it is just possible that a projecting tail end could interfere with winding. Also, the knots may retain traces of cleaning solvents or moisture, either of which may corrode the drum.

While the gear is off the drum, test the click for play. Some clicks resemble a bent T, as they have a tail for ease of letdown; others are without a tail, being of a simple L shape. In both cases, the vertical stroke of the letter-shape passes through the no. 1 gear edgewise, and forms the arbor of the click. It is important to realize that riveting is not involved; the arbor is threaded, and a loose click must not be hammered in an attempt to tighten it.

If the click is unduly loose, ease the click spring aside and unscrew it completely. As both threads will be badly worn, they must be renewed and the click rebuilt. Cut off the threaded arbor, file the click flat, and drill a hole through it where the arbor was rooted. The hole should be lightly countersunk on both sides, and its diameter should take up half (or a trifle more) of the available width of metal.

Now tap out the hole in the no. 1 gear, using the smallest tap that will cut sound threads. Try to use a thread which has close to the same number of threads per inch as the original, for this will give the best results. It does not matter whether an American, English, or metric thread is used, but it is essential to tap the hole square with the gear. Using a drill press by hand is one way.

Select a short steel screw of the chosen thread, cut off the head, and, on the lathe, turn down a short section of the screw until it is a firm fit into the hole in the click. Press it home in the vise, using a small nut or spacer to clear the emerging end, and then file the end down almost to the click, leaving enough for riveting. Ease the click spring away and screw the click in place. Turn the click against the threads, unscrew it, and then tap it on again, repeating the process until the threads synchronize with the click when it is in the operating position and is close to the gear at the same time. This is necessary because, as the threaded arbor rotates, it also moves axially; hence, without synchronization the click may operate with a gap between it and the gear, with a consequently poor mesh.

When the correct orientation has been found, rivet the arbor firmly, and saw off the excess threads. (Do not use cutters, as they flatten the threads.) File the end smooth and screw the click in place. See that the click spring is still tight on the no. 1 gear and that it exerts adequate pressure.

Mount the arbors in turn in the lathe, without their gears, and polish both pivots. Now is also a good time to try the winding crank on the squares, and to file out any irregularities on them, bringing them to a common size. Supply a smaller-sized crank if necessary, or use the **key shrinker** to obtain a good fit.

Drill out the remains of the friction-collet retaining pin, where necessary, so that the hole is clear. Now run a no. 50 drill through it, and then tap the hole to size 2-56 N.C. Drill out the hole in the friction collet with a no. 30 drill, followed by a no. 29 drill; this will give a snug clearance fit for the head of a no. 2 cheesehead screw. An English or metric screw of similar size may also be used, provided that the hole sizes are changed accordingly. Make sure that the screw does not project through the gear and foul the drum; shorten the screw with a file if necessary.

Increase the dishing of the friction collet until there is no play between the no. 1 gear and the drum. This is done along the same lines as shown for a hand collet. (See **Hand collet** in Part IIII.) When adjusted correctly, the arbor must not be stiff to turn; it should turn easily in the fingers, but without any suggestion of spinning.

The reasons for fitting a screw instead of a pin should be noted: The screw can be tightened down fully to hold the collet in position, without influencing the collet pressure, and can be easily removed for subsequent overhauls.

Treat both no. 1 arbor assemblies alike, as described above, according to condition, and reassemble them.

Step 27 deals with the pivots. Since in some cases they may have been hardened, they are not as easy to deal with as those of the ogee and other American clocks. (See **Pivots** in Part IIII.)

Step 28, bushing, follows logically after the pivots have been trued and polished. As rolled brass did not come into general use until about 1825, the plates of movements made prior to that date were cast and filed, and hence are somewhat uneven in thickness. These old plates are up to 0.130″ thick, and as variations of some 0.030″ are common, the length of bushings changes with location. It is therefore best to cut them a little longer than usual. Note that it is occasionally necessary to leave a bushing slightly long on the inside, to take up excessive end play. (See **Bushing** in Part IIII.)

Continue with *Step 29,* pegging out the pivot holes, and with *Step 31,* checking the **meshing.**

Step 32. Fortunately, cast brass, although much easier to bend, does not warp as readily as rolled brass, and so for the most part variations in thickness constitute most of the aberration likely to be found. What warps there are can be taken out. The pillars, however, must not be used as levers; always use a vise, and protect the plates from its jaws with chops of wood, plastic, or other suitable material. Somewhat greater force is required for the later rolled-brass plates, but in either case the required movement is usually quite small. It is important to note that some curves found in plates are intentional, serving to take up arbor end play.

Step 35, dealing with the hammer spring, is next, although it is feasible to do this after the movement is assembled, when the spring is then adjusted with a pair of screwdrivers.

Screw the spring in place on the back plate, and fit the plates together with only the hammer arbor between them. Try out the hammer action and adjust the spring as necessary to obtain consistent results. Residual spring pressure must be held to a minimum or the train may refuse to run reliably. The spring can be adjusted while in position, by using two horizontally held screwdrivers, one from each side. One is applied to the center of the spring, the other to the top; the spring will bend accordingly when pressure is applied.

If the hammer head is loose, it can be soft soldered on again only if the hole is deep enough—say, 3/8" or more. If the hole is shallower than this, soft solder may give way; in this case is better to silver solder it.

Step 36. Examine the hour pipe, and see that it is clean inside and out, not weakened, or loose in the bridge. This component is not as strong as it appears at first glance. Set the minute arbor in its pivot holes, pin the plates together, and slide the minute pipe onto the minute arbor. Put the bridge in place, in its correct sense, and set the hour gear on the hour pipe. Now spin the minute arbor, and note the concentricity: The minute arbor must be straight, and there should be no binding between it and the pipes during rotation.

The freedom of these components relates to Step 23, for unless the plates are square with each other the hour pipe will not be concentric with the minute pipe. It might then seem desirable to bend the bridge to obtain clearance; however, this would be a wrong move. It is best to go through Step 23 again. If eccentricity still exists, the hour pipe must then be moved slightly. This is done by bending the bridge itself a little; *never attempt to bend the hour pipe,* for it will simply fracture at the bridge. Even the bridge itself is weak in the center, and must be adjusted with great care. (If any locating pins are present, they must be removed by hammering them out towards the front of the bridge, and then returned to their positions, if possible, after adjustments are complete.) It is feasible to raise either end of the bridge a trifle, through careful hammering and bending. A small sideways tilt can be attained by clamping each end of the bridge in turn in the vise and torquing the center section with a small adjustable wrench. Note that this action takes place on the same side of the hour pipe—that is, the wrench is always between the hour pipe and the vise. When concentricity is attained, the feet of the bridge must be in a common plane, so that the bridge cannot rock on the front plate. Run a file across both feet at once, as a final check on their alignment and flatness.

Continue with *Step 37,* final inspection, and *Step 38,* reassembly. In positioning the movement for reassembly, even though the rear plate will lie reasonably flat on the bench, it is best to use an assembly box to keep dirt out of the pivot holes. The hammer spring may be fitted before or after the arbors, according to choice.

As there are no lifters between the plates, the movement goes together quite easily; no attempt need be made to synchronize the strike train. Leave out the pallet arbor for the time being, and pin the plates together temporarily. Fit the center spring in place, with its convex side nearest to the front plate, and add the cannon pinion and bridge. See that the tapered steel arbors that support the reduction gear, the lifter, the rack hook, and the rack are firm in the front plate, and then fit these parts in place in that order; pin all save the reduction gear. It may be necessary to add the hour gear assembly simultaneously with the rack, though this is not always the case. Adjust the rack spring so that it exerts only a small force when the rack is against the rack stop pin. It is best to bend the tip of the spring into a V so that it is unlikely to disengage. To maintain good access, do not fit the bell post and bell for the time being.

Step 39. Check the end play and meshing, as before; the amount of end play varies somewhat due to variation in plate thickness, and backlash is usually considerable, owing to wear on the pinion leaves. Little can be done about this, but it does not pose a problem unless it is severe enough to stop the clock, in which case new pinions must be cut. As this often involves making new arbors as well, it is a long and expensive job which very few people care to undertake.

As a rule of thumb, *minimum backlash all around* seems the only practical criterion for clock gearing. It is rather pointless to discuss pitch diameters and so on, for such parameters are generally unknown and not readily measurable. Also, studying a drawing of a correctly meshed gear and cut pinion cannot be followed up by a visual end-view check. Recognizing and obtaining optimum mesh depends largely upon experience.

Step 41 involves setting up the strike train. Fit the minute hand on its arbor and turn it to twelve o'clock. Ease the reduction gear forward so that it no longer meshes with the cannon pinion, and rotate it until the unlocking pin allows the tail of the lifter to fall. Push the gear back into mesh again, and check that the minute hand is vertical when the drop-off occurs. It may be necessary to repeat this, moving the meshing a tooth either way. If insufficient accuracy is obtained, put the minute hand in one of its three other positions and repeat the procedure. There is no need of a dial for this; the eye is perfectly adequate.

Pin the reduction gear in place when it has been thus synchronized with the cannon pinion. The importance of straightening the minute hand, done in Step 25, will now be evident.

With the minute hand vertical, bring the hour gear and snail assembly forward until it unmeshes from the reduction pinion. Lift the rack hook, and turn the snail until the pin on the rack tail contacts the center of the twelve step. Maintain this position and slide the hour gear back into mesh.

Fit the gathering pallet lightly on its tapered square arbor (see Fig. 26). Then run the strike train by hand, checking that it strikes twelve o'clock correctly. If the gathering pallet lands on top of a rack tooth, a slight lengthwise twist of the rack tail will usually correct things. If the count is more or less than twelve, however, the angle between the rack and its tail must be changed. As the tail is often soft soldered on its pipe, heat must be applied before the angle can be corrected. It may be necessary to repeat this adjustment a time or two, because this angle is critical. Unless accuracy is achieved, the clock will strike erratically or not at all. When the counting is satisfactory, pin the rack on its arbor. An S-shaped wire is as good as or better than a tapered pin for securing the four strike-train components on their arbors.

Pin the minute hand on with its collet and check the counting action around the twelve positions; when twelve o'clock and one o'clock are set up correctly, the rest usually fall in line, unless the rack teeth are erratically spaced. If a problem arises, do not be tempted into filing the steps of the snail, for they are almost never at fault.

Twist the lifter tail until it allows the unlocking pin to slide under it when the minute hand is turned backwards. There are bevels on the twelve o'clock–one o'clock step on the snail, and on the rack-tail pin, to prevent damage should the hands be turned backwards. Unfortunately, these measures are not always effective, for if the rack tail does not distort, its pin will score the snail. For these reasons it is not advisable to turn the hands backwards on this movement, certainly not past the twelve o'clock mark. However, when the lifter tail has been twisted, as described above, and the action is visible during setting up, it is very useful to rock the minute hand back and forth; as long as the rack is moved when necessary, no harm is done.

When the rack is fully counted, the tail of the gathering pallet hits the rack pin and stops the strike train. In this position the hammer tail should have just dropped off one of the lifting pins on the S2 gear. If this is not the case, pull the gathering pallet off and refit it in another of its four possible positions, continuing this procedure until the best one is found. Remeshing the S2 gear with the S3 pinion gives a finer adjustment, but this is seldom necessary.

When striking is over, and the hammer tail is free of the lifting pins, the warning pin on the S4 gear must be close to the top of its travel—that is, at twelve o'clock. If it is not in this position, take out the top two pillar pins and pry the plates apart far enough to remesh the S4 pinion with the S3 gear so that the warning pin is at the top.

Bring the minute hand to the 45-minute mark, and then apply finger pressure to the strike train. Slowly advance the hand until warning occurs; if the hammer moves at all, repeat the last two procedures until it remains at rest during the warning. Push the hand up to the 60-minute mark and allow striking to occur; continue around the dial, counting each hour in turn until all the steps in the snail have been checked. If all is well, rest the end of the rear pivot of the S3 arbor on an anvil or vise, and gently tap the gathering pallet on firmly, using a hollow punch. Insert the tapered pin, if a hole is provided for it. Usually, however, no such hole exists, in which case a small set-screw collet is a good way to secure a pallet with a tendency to loosen.

Although there are several adjustments on the front plate, and the S2, S3, and S4 arbors have to be synchronized, the longcase striking mechanism is very reliable when set up along the lines indicated above. Now remove the minute hand, but replace its collet and pin. This will prevent the hour gear assembly from slipping out of mesh, and will also free the minute pipe from the pressure of the center spring.

Step 42. Set up the escapement as described under **Escapement** ("Recoil—Solid Pallet—English Longcase") in Part IIII.

Step 43. With the escapement in place, apply firm pressure with thumb and fingers to the T1 gear and check that the train will run. Note that it may be reluctant to do so if the center spring is pressing the cannon pinion against the bridge; this is why the collet and pin were replaced after the minute hand was removed in Step 41.

Step 44. In replacing the cables, you should treat both drums alike; it does not matter which is dealt with first. Turn the drum so that the cable hole faces outward and examine the edge of the hole; generally it is quite sharp, sometimes sharp enough to cut part way into the cable. This is a point generally ignored by makers and repairers alike, yet easy to remedy: Take a Swiss rattail file—preferably a broken one, for a larger end diameter—and apply its end to the hole so that its side rests on that part of the lip on which the cable lies and its end abuts the other side of the hole. Now rotate the file, whichever way cuts best, until the sharp rim of the hole assumes a small radius. Because the hole is at the end of the drum, the usual countersink cannot be used.

See that the cable-alignment pin slopes towards and touches either the ratchet gear or the front end cover of the drum, whichever it is placed against. Most clocks wind from the rear, the ratchet-gear end, but some wind from the front of the drum; the alignment pin is placed accordingly.

Some drums have spiral grooves cut in them to accommodate the cable, while other drums are smooth; in either case, about ten feet of cable should be allowed to fill them. Any excess can be cut off later. The cable used may be of gut, braided nylon, or monofilament nylon fishing line, the breaking strength being about 100 pounds in all cases. This may seem unduly high, as a pull of only 6 or 7 pounds is required to drive the clock; however, wear, deterioration, and stretching all have to be taken into account. Also, if the clock is carelessly wound and the pulley slams into the seatboard, the strain on the cable escalates to several times the required pull, so a wide safety margin is needed.

The 12–14-pound weights of a longcase are compounded, once, by their pulleys. This gives a 6–7-pound pull on the seatboard and on the clock drum. The weights fall about 4½′ as they pull some 9 feet of cable off the drums.

The movement and dial weigh about 8 pounds, the weights around 28 pounds, and the bob some 2 or 3 pounds. As all this rests on the seatboard, this old, battered, cut, and weakened piece of wood must support in the area of 40 pounds total weight. In view of this, it is important to inspect it to see that it is strong enough to perform its task.

Pass the end of the cable through the drum's cable hole, and draw it out through the larger hole in the end of the drum with a small hook of steel wire. Tie a tight knot in the end of the cable and cut off the excess, leaving a tail about ¼″ long; then pull the knot inside the drum. Give a strong radial pull to tighten the knot and to check that it will not pull through the cable hole. Do not wind the cables on as yet, but coil them on the bench; trailing cables invite disaster.

Step 45. Oil the movement along the lines indicated in **Oiling** (Part IIII). Fit the bell and bell post to the back plate, and then rotate the bell or bend the hammer arm until a clean, sharp strike is obtained.

Step 46. Examine both pulleys for chipped or bent rims and excessive wear. Kinks in the rim may fray the cable, and they must be straightened out with a pair of pliers. A deep chip usually entails making a new pulley.

The center hole in the pulley is often badly worn, and this sometimes results in a sideways slew which causes the rim to bind against the U-shaped steel yoke on which the weight hangs. This action wears one side of the rim to a sharp edge, and a new pulley, arbor, and yoke are then indicated. If the wear is less severe, bend the yoke in the vise until the rim of the pulley runs centrally.

Pulleys can be bushed, of course, but because of the heavy one-sided pull, a new arbor should be made at the same time. When the pulleys are satisfactory do not forget to oil them. Judging from the amount of wear usually present, this has obviously been overlooked at times.

Step 47. Set the movement in position on the seatboard, checking that the cable will not foul the edges of the rectangular clearance holes. Clean them out with a wood chisel or a coarse file if necessary. Screw the movement to the seatboard, using washers underneath the screws to prevent their heads from biting into the wood. The screw heads should have a low profile, so that the pulleys do not strike them when the clock is fully wound. This kind of contact is responsible for kinked pulley rims.

The procedure for cleaning, restoring, or repairing the dial is explained in Chapter 8 under "Longcase White Dial Repairs." When the necessary work has been done, pin the dial in place and replace the hands. Look closely at the second hand and the hour arbor where they pass through the dial, making sure that both have adequate clearance.

The hour hand may be on a clutch or may be fixed, but it must line up with the hour set by the position of the snail. Make sure that the minute hand is vertical when the strike is actuated. When pushing against the minute-hand collet to insert the pin, it is easy to damage the crutch; leaning the movement away from you until the pendulum cock rests on a block of wood will obviate any chance of damage and protect the crutch.

If the clock case has not been restored, it is now time to leave the movement and deal with the case. Various aspects of this are discussed in Chapter 9.

When the case has been restored to soundness, set it up vertically, using a spirit level. Make sure that it cannot sway, and then set the movement in place.

Turn the drums so that their cable holes are directly over the rectangular holes in the seatboard, and feed the cables through to hang down inside the case. Fit the pulleys on, and take any twists out of the cable.

Push the free end of the cables through the appropriate holes in the seatboard; take up enough slack to raise the pulleys about 10" above the floor. This is a little less than the height they will assume when the weights run down.

Withdraw each cable in turn and knot a small loop at the point where it passed through the seatboard. Make the loop about ¼" longer than the thickness of the seatboard. (Do not cut the free end off yet.) Push the loop through the hole used previously, and put a smooth finishing nail, not less than ⅛" thick, through the loop. Before a gut cable is knotted it should be softened in water, or it may fracture.

Hang the weights and wind the clock up slowly, watching to see that the cable winds evenly onto the drums. The forward and backwards tilt of the clock governs the wind-on, and it may be necessary to shim the seatboard at the front or back, on both sides, until even winding ensues. If the drums are plain, all the cable will be accepted; in contrast, a sprial-cut drum may fill up before the pulley reaches the seatboard. Adjust the position of the loop accordingly, and cut off excess cable.

If used on a permanent basis, the finishing nail used with the loop above might possibly work its way out of the loop, thus allowing the weight to fall. As the resulting crash and damage is unlikely to enhance horological morale, a better way must be found. Two very sound methods are to bend down one end of the nail and push it into a second hole in the seatboard, or to make a neat figure eight from a discarded end of solid electrical wire. For the latter, no. 14 copper wire is ideal, for it is easy to bend and its plastic coat is kind to the cable.

There is little chance of the loop's slipping from the waist of the figure eight, as the ends of the wire meet at the top or bottom of the figure—not in the middle. Long-nosed pliers will be needed when securing the loop under the movement. The advantages of using a loop are that it avoids a big, ugly knot which will have to be untied at some future date, possibly by you, and that it makes it easy to separate the movement from the seatboard for future servicing. Keep the loop as small as possible, however, so that its knot, under the seatboard, will not encroach on the pulley groove—it will not run in it.

Step 48. As the rack-and-snail strike mechanism is virtually self-correcting, the strike does not go out of synchronism unless the clock is allowed to run completely down, and the strike side stops before the time side. When this happens, wind the strike side up and allow it to strike, and all will be in order again. If the clock does not strike, it will probably be because the rack-tail pin has overridden the snail, when the twelve o'clock count failed to occur. In this case, reach behind the dial at about the ten o'clock level, and push the rack inward by hand until its tail clears the snail, thus allowing the clock to strike.

If the hour hand is on a clutch, it is possible for it to be displaced during winding, but it is easily returned to its original position.

Aside from these readily corrected situations, rectifying faulty striking will usually entail removing the dial to see what is amiss. The most likely fault occurs when the rack spring is set up too strongly. As the rack is repeatedly slammed back in normal use, the angle between it and the rack tail widens slightly or the tail twists a little, giving the same effect. The gathering pallet can then no longer mesh correctly with the rack, and instead lands on the tip of a tooth, thus jamming the strike train. The remedy here is to adjust the rack-to-rack tail angle a little, as previously explained in Step 41, and then weaken the rack spring slightly to prevent a recurrence of the trouble.

Synchronizing the usual ordinary simple calendar can be done at any time of the day or night—except during two periods every twenty-four hours when the calendar pin is engaged with the teeth of the sixty-two-tooth calendar gear. Each engagement lasts about two hours, and will be felt as a solid resistance if adjustment is attempted during these two periods.

Whether the calendar appears as a hand akin to the second hand, or has an aperture, it is best to set it by opening the case door—not the hood door—and reaching up behind the dial to turn the gear with a finger. Pushing the hand itself, or poking a finger in the aperture, will dirty the dial and slowly wear away the figures. Never force either type of calendar gear to move.

Step 49. Before installing the pendulum, study **Pendulums** (Part IIII), for it is quite likely that repairs will be necessary. Despite its apparent simplicity, the pendulum of a long-case clock can cause a great deal of time-consuming trouble if it is not given the full treatment. We cannot dictate to a pendulum; it must have "dominion over the clock," as the old-timers used to put it.

When the pendulum is judged to be sound, raise the suspension spring up through the crutch and slide it into the slit in the pendulum cock, so that the chops fit snugly in the recess provided for them.

The junction between the suspension spring and the pendulum rod is usually a small rectangular block of brass, which is a free fit in the rectangular crutch hole. First impressions of this arrangement easily lead to the all-too-common belief that the crutch, in addition to driving the pendulum, keeps it swinging truly and stops it from wobbling. The true state of affairs is that if the pendulum insists on wobbling, and the crutch tries to prevent it, the resulting friction either stops the clock or causes an erratic and much reduced swing which results in poor timekeeping. This is simply a manifestation of the concept that if one meddles with a governor, there is no point in having a governor.

Looking at the movement from above, check that the crutch is parallel to the slit in the pendulum cock, and that both are at a right angle to the plane of the back plate of the movement. Note that the slit is the critical item here, rather than the cock itself, for quite often the slit is poorly cut and not parallel to the sides of the cock. Adjust both the cock and the crutch accordingly.

Check that the dial is vertical. It will not be far off if the cables wind on evenly, but a small adjustment to the seatboard is permissible if it does not upset the wind-on. If necessary, ease the crutch towards or away from the back plate, keeping it level, until the pendulum block lies centrally in it; it must not touch at either end.

Now set the pendulum swinging gently, and note the beat. If the escapement jams, then the weight of the pendulum has lowered the cock very slightly. The cock must be moved up a trifle to compensate. If the beat is uneven—and it usually is—the crutch wire must be bent to the left or right. Simply pulling or pushing the crutch with one hand is asking for trouble in the form of bent escape-gear teeth. So, to prevent damage, use two hands.

Operate the pendulum by hand on both sides of the center line, and note on which side an even beat can be obtained. The crutch wire will require bending from this side towards the center line. To do this, stand in front of the clock and put both of your hands behind the movement. If the crutch wire needs to be moved to the right, put the tip of the second finger of your right hand midway up the wire. Spread the first and second fingers of your left hand and rest their tips against the top and bottom of the wire respectively—not against the crutch itself. Pushing your hands together in this position will bend the wire to the right. Some wires are quite stiff, and if appreciable force is required it is a good idea to use two fingers close together in the center. To move the crutch to the left, repeat this procedure with the hands reversed. These adjustments may require repetitions until an even beat results.

Bear in mind that consecutive ticks are of unequal loudness—do not allow this to influence the process of putting the clock in beat. A clock tilted to one side will sound thus: "Tick-tock—tick-tock—tick-tock." If tilted to the other side: "—Tock-tick—tock-tick—tock-tick—." A clock which is in beat gives out an even "Tick–tock–tick–tock–tick–tock;" this is the desired rhythm.

When the clock is properly in beat, observe the pendulum's swing and note any aberrations such as a bowed swing, wobbling, a twitch at the end of the swing, and so on. If such faults have not disappeared in about ten minutes, in favor of a steady, in-line swing, stop the clock and center the pendulum.

Check that the jaws of the pendulum cock are close enough together, so that the suspension spring is not loose in them. Check that the chops at the top of the spring are equal in length; one side must not take all the weight. See that the spring is firm but not tight in the block.

Next, look down through the case door at the bob, and gently rotate the pendulum shaft, first one way, then the other, as far as the crutch allows. Set the shaft at these two limits, and release it; it should turn back from both extremes and assume a position favoring neither. If it does not, then the suspension spring is slightly twisted. To remedy this, apply a firm twist with a thumb and finger to the top of the suspension spring just under the pendulum cock, to turn the pendulum away from the position it favors. Repeat this until the pendulum no longer binds on either side of the crutch.

In most cases these measures will cure a wobbling pendulum; if the fault still persists, renew the suspension spring and try again. Contrary to what might be expected, a narrow suspension spring works better than a broad one. In addition, curiously enough, a battered spring may swing truly and a flat, perfect-seeming spring prove useless.

While the pendulum shaft should be reasonably straight, do not waste time bending it in an attempt to cure wobbling. The trouble always lies at the top of the pendulum, in its suspension.

Step 50. Turn the minute hand around the dial a few times—always allowing the clock to strike—and adjust the hands, if necessary, so that they do not bind against each other. Make sure that the hour hand cannot drop and so foul the second hand or calendar hand.

Put the hood in place and close its door. Check that neither the tip of the minute hand nor the end of its arbor is in contact with the glass.

Set the clock to the correct time, and after half an hour or so measure the arc of the pendulum just below the tip of the rating screw. This figure varies somewhat, 6–8″ being usual after an overhaul. I consider 5″ to be a little low, but the arc tends to increase slightly after a week or so, when the clock settles down, so measure it again later.

Some of these old longcase clocks have "recorded" two centuries. If they are given an overhaul such as that described here about every five years, they are good for yet another century. I should certainly expect them to continue ticking long after I have stopped.

CHAPTER

6

French Clocks

French clock cases vary widely in appearance, but the movements they contain differ very little, as they are for the most part merely variations on a common theme. It does not matter whether the case is a small, fairly simple, highly polished wooden affair, a massive marble edifice weighing 80 pounds, or a detailed casting of a horse or a human figure—the same basic round movement will be found in all. Wood, zinc, brass, glass, cast iron, marble, porcelain, and other materials were widely used to make cases which have ornate embellishment as their chief characteristic. Simplicity of design seems to have been unpopular. A great deal of work and skill went into making such cases.

It seems only natural that the same love of detail and accuracy of execution should be displayed in the ubiquitous round clock movements which were made in great numbers, with only a few minor design preferences of individual makers as distinguishing features. Although dimensions may differ somewhat—and the plates may occasionally be square instead of round—the marked similarity of nearly all French movements is very obvious.

a.

Fig. 27. French Clocks
  a. Wooden Case
  b. Marble Case
  c. Ormolu
  d. Crystal Regulator

b.

c.

d.

8-DAY SPRING MOVEMENT

Movement: Fig. 28 Cases: Fig. 27a, b, c, d
Key size: 5 & 6

Both the recoil escapement and the Brocot (pin pallet) escapement are used in the round French movement, the former being the most common. Aside from the method of locking, the strike train operates rather like that of the English longcase movement; however, in every other way the differences are very marked. While the English longcase and most American clocks have what are probably the thickest pivots and arbors to be found in the world's common domestic clocks, the French movement has the thinnest.

Despite the fact that they were made in great numbers, there are no flashings or other evidence of mass production on any of these finely made movements. Their substantial circular plates, some 3–3½″ in diameter and from about 0.080″ to 0.090″ in thickness, are usually free from blemishes. Cut pinions are used throughout, and both they and their arbors are hardened and polished to a high degree. The gearing is finer than in most clocks, and the mainsprings are less powerful than those of the American and English clocks.

The maker's name is sometimes engraved on the back plate, often with a stamped-in indication of an award for excellence. There is certainly not the slightest evidence of cost cutting; there are no bent sheet-metal parts, and all the cocks, together with the three or four screwed-on subplates, are machined from solid metal. The rack hook is curved, as are the tails of the clicks. Steel items such as screws and the hammer arbor are polished and heat blued, though this has often worn thin over the years. Fine regulation is accomplished through the dial via a silk-cord suspension in the earlier movements, but the majority of clocks use the later, rather involved, Brocot mechanism. Undoubtedly many cost-saving changes could have been made, and the whole movement is possibly somewhat overengineered.

French clockmakers do not seem to have adopted the fusee as enthusiastically as their English counterparts, but French clocks are good timekeepers just the same. They are, however, much less robust than, say, American clocks, and the position in which they will run is more critical. Owing to the usual method of mounting, it is easy to accidentally rotate the movement slightly during normal winding, and this small shift, which may well go unnoticed, usually puts the clock far enough out of beat for it to stop.

The movement is pinned to the dial by means of three pillars. The dial assembly, which is solidly made and relatively heavy, sits in a hole in the case front and is tensioned to the case back by two straps, one at each side. Although this is a convenient arrangement, it is not the best—for the above-mentioned reason (rotation during winding) and because if the case is wood any relaxation in the wood will loosen the movement.

First, open the rear door of the clock and carefully unhook the pendulum; depending upon access, this may or may not necessitate first removing the bell from its bracket. If a gong is used, access is usually adequate, though somewhat cramped. On either side of the rear door are the two screws that tension the straps; take out these screws and then remove the bezel, dial, and movement from the front of the case, as a unit, taking care that it does not fall out prematurely. Open the front bezel and remove the hands (see Fig. 3). Take off the dial by unpinning its pillars.

These actions correspond to the first five steps of the ogee procedure; as the layout of the clock dictates that they cannot be done in the order of the ogee, I have not numbered them separately.

Continue with *Step 6,* if required. However, as the cases of French clocks are tolerably well closed up, these movements are rather less likely than most to need a primary rinse with a general purpose solvent.

Step 7. See that the movement is complete, and check for serious damage. As the gears do not project much beyond the limits of the plates, their teeth are to some extent protected from accidental damage. I have noticed that these movements do not need bushing as often as American and English types; this is because the hardened, highly polished, and relatively tiny pivots require less power to drive them. At the same time, I consider the mainspring barrels to be a trifle on the thin side and their teeth a shade too fine.

One of the few variations found among these movements lies in the strike train; some use the count-wheel system and others the rack and snail. As the latter are somewhat more numerous, it seems best to consider our present movement as of that type.

Step 8. Because the finish on the plates is above average, scratches show easily, so it is better to jot down the locations of those pivot holes that require attention; this is preferable to marking the plates.

Step 9. Note the positions of strike-train gears when striking is completed. The present movement is one of those, referred to in Chapter 2, that has pins on two of the strike gears; the rest positions of both these pins should be noted before disassembly. Check the operation of the strike train to see that it works correctly; although it is unlikely to be wrongly set up, there is no point in recording settings that are

Fig. 28. 8-Day Spring Movement

incorrect. As always, the hammer must not be raised at all except when striking is actually taking place.

Step 10. There is little need to mark any components of the gear trains save for the usual small "S" on the strike barrel, which should be done neatly and close to the hook for ease of identification. Arbors S4 and S5 of the six-arbor strike train are rather similar, as both carry a gear with a pin on it; however, the S4 gear is larger and closer to the source of power, S1. In the time train no ambiguity is possible.

It is possible to be confused by the small subplates, of which two are usually on the back plate and one on the front, but their dowels generally prevent wrong reassembly.

Take off the reduction-gear cock and the gear itself, together with the snail and hour gear. Ease the minute pipe and cannon pinion off the minute arbor. Unpin and remove the rack hook, lifter, and rack from the front plate, and the hammer and pendulum cock assembly from the back plate.

Step 11. Let the mainsprings down fully, using a letdown key; the ornate tails on the clicks facilitate their release.

Step 12. Unscrew the pillar screw at the top of the movement; this is normally hidden by the pendulum cock. Take out the three pins that hold the remaining pillars, carefully pry the plates apart, and then remove all the arbors. Remove the ratchet-gear retainers, the clicks and click springs, the subplates just mentioned in Step 10, and the gathering pallet. Great care must be taken in this operation, for the arbor is delicate and easily broken. Thread a fine wire through the hole in the gathering pallet and wire the pallet to a gear so that it does not get lost. Unscrew the hammer spring and the rack-hook spring when possible.

Step 13. Scribe a small "S" on the strike barrel cover and on the strike arbor. (Either end of the arbor, or one of the sides of the winding square, is a good spot to mark.) Pry off both barrel covers and remove both mainsprings; see **Mainsprings in Barrels** in Part IIII. Scribe another small "S" on the soft outer end of the strike mainspring. (It is never safe to assume that the two mainsprings are alike and the therefore interchangeable in any clock; often such is not the case.) Check the amount of play between the no. 1 arbors and their respective barrels and covers. (See **Bushings—Barrels** in Part IIII if bushing is required.)

Step 15. Put all components through the cleaning process, leaving out the hammer only if it has a leather insert.

Step 16. Inspect all components carefully, including the teeth of the rack and all gears. See that no dirt remains between pinion leaves or gear teeth.

Step 18. See that all gears and collets are firm on their arbors; check the same points mentioned in Step 18 of the English longcase text (Chapter 5, page 74.)

Step 19. Make sure that the fly is in good order and tensioned correctly on its arbor.

Step 20. Test these pins to make sure that they are straight and firm: the rack-stop pin (which limits the rack's travel), the hammer-lifting pins on the S3 gear, the locking pin on the S4 gear, the warning pin on the S5 gear, and, on the backplate, the crutch-wire-stop pins and the hammer-stop pin.

Step 21. Check the Brocot pendulum suspension and fine-adjustment device that constitute the pendulum cock. Like the other cocks on this movement, this mechanism is milled out of solid brass, and no pains were spared in the making of it. It works well, but requires its own particular double suspension spring, a rather long and narrow version. The upper chops of this spring have a notch in them, as have its modern replacements. As the notch seems to serve no mechanical purpose, I assume that it is there as a mark to ensure consistent assembly.

Step 23. See that the pillars are firm in the front plate and line up with the holes in the back plate.

Step 24. Remove any rust, as for the English longcase movement (page 74).

Step 25. Deal with the hands as in the ogee text.

Step 26. Examine the clicks to make sure that they are sharp and true; see that they mesh well with their ratchet gears. Check that the teeth of the ratchet gears are sound. Bend the click springs a little, if necessary, so that they exert adequate pressure.

Continue with *Steps 27* and *28,* pivoting and bushing respectively, referring to **Pivots** and **Bushings** in Part IIII.

Step 29. Peg out all pivot holes with a piece of small-diameter pegwood, as described under **Pegging Out** in Part IIII. Since the holes here are smaller than usual, the tip of the wood may break off; if this happens, extract the broken tip with pliers, if possible; if this is not feasible, push the broken end out with more pegwood. If steel wire has to be used, see that there is no burr on the end which could score the hole.

Step 31. Test the meshing of the gears on any

arbors that may have been disturbed by the bushing in Step 28.

Step 32. Check the plates for flatness. They may vary a little in thickness, but warps are unlikely to be troublesome in these movements since they are relatively small and sturdily made.

Step 33. Fit the cannon pinion and minute pipe on the minute arbor, and check that its friction is adequate. If the pipe is too loose, squeeze the sides of the cutaway section very slightly together with a pair of pliers; it is very easy to apply too much pressure in this adjustment, as only a few thousandths of an inch are required.

Step 35. Inspect the hammer. Although the head does not often loosen, the lower end of the hammer arm is sometimes loose in its collet, for the hole is not really deep enough. The hammer can be rebuilt with a new and slightly larger arm, but this too can work loose in time, as the hole cannot be deepened because of the hammer arbor. It is feasible to drill a new, deeper hole passing above the arbor, but it is not easy to obtain a tight fit in it. As few people wish to make a new and larger collet to correct the problem, it makes sense to simply clean everything carefully and soft solder the arm securely in place. If this is done well there need be little evidence to offend the purists except a small, neat fillet of solder around the base of the hammer arm.

Step 36. Clean out the hour pipe, and ensure that it turns freely on the minute pipe. Try it at several depths, for a small bind is not always apparent in all positions.

Step 37. Give all parts a final inspection, bearing in mind that it does not take much dirt or distortion to jam the fine gears of these movements. Then prepare for reassembly.

Step 38. The T4F pivot usually has its own subplate on the front plate of the movement. Screw this subplate in position, and lay the front plate on the assembly box. If the box is too big or somehow does not fit, screw three pillars or spacers about 1″ long into the dial pillar holes, and use them as temporary supports.

Starting with the minute arbor, put the arbors in place in the front plate and close up the back plate. Apply only light pressure for this, as the fine pivots are easily broken; the clearance holes at T3B and S3B make the job easier. Pin the three pillars lightly, and screw the two subplates on the back plate; do not tighten down the subplate at S3B yet, as it may have to be removed again in Step 41 below. Screw in the rack-hook spring and hammer spring, and fit all the front plate parts except the reduction gear.

As the unlocking pins are on the cannon pinion, there is no need to synchronize the reduction gear as in the English longcase movement. Fit the hammer on its arbor.

Step 39. Test all arbors for end play and all gears and pinions for backlash and freedom from binding. (See Step 39 of the English longcase movement, page 78.)

Step 41 is setting up the strike. Fit the minute hand on its square and turn it to actuate the lifter. There are two unlocking pins, 180° apart, on the cannon pinion, but as they are not equidistant from the center of rotation they raise the lifter to two different levels. The innermost pin raises the lifter only enough to unlock the warning pin, without raising the rack hook enough to allow the rack to fall. When the lifter falls, the train is then free to count only one tooth of the rack, thus marking the half hour. Reposition the minute hand as necessary to coincide with this striking pattern. With the hand at twelve o'clock, lift the rack hook and turn the snail until the rack tail lands on the twelve o'clock position at the bottom of the step in the snail. With the snail held thus, fit the reduction gear and screw down its cock.

Actually, there are no specific steps on the snail, but long use has usually caused the end of the rack tail to make twelve small indents in the edge of the snail. Check now that the rack tail lands on these marks as in the past, and if necessary adjust the reduction-gear meshing to achieve it. Unlike the English longcase movement, the French movement does not allow the rack tail to override the snail if the clock hands are turned backwards or the clock continues to run after failing to strike. Consequently, strike failure will cause the time train to stop, at about 12:30, when the step in the snail jams against the rack tail. A fault in the strike train cannot therefore be ignored, as it might be in other clock movements. If immediate repair is not feasible, the rack can be wired to its stop pin as a temporary expedient allowing the clock to run. I cannot recommend this measure, but it is vastly better than removing the rack or some other part, as happens all too often. Such parts invariably get lost, and the clock descends into the basket-case category. Obviously, it is much better to effect an immediate repair.

Set the movement down on the bench in the upright position, as though it were in its case, with the front plate to the right. Lift the rack hook, thus allowing the rack to fall. Rest a finger on the S3 gear and gently push upwards so that it rotates clockwise as its pins operate the hammer. When striking ceases, the locking pin on the S4 gear will be stopped against the top surface of the locking hook, at about eleven o'clock. If necessary, adjust the hammer spring and the rack-hook-cum-locking-hook spring by bending them slightly at a point about ¼" from the front plate; a weak pressure is all that is required.

In the position reached above, the hammer tail should have just dropped off one of the pins on the S3 gear, and the hammer must be entirely at rest. If the hammer is raised, remove the S3B subplate and remesh the S3 pinion; repeat, if necessary, until the hammer is free. When the train is locked as intended, the gathering pallet should be at about ten o'clock, and the warning pin on the S5 gear at about nine o'clock. The gathering pallet is easily removed and repositioned, but if the warning pin is out of position the plates must be gently eased apart and the S5 pinion remeshed with the S4 gear, as there is no convenient subplate on the S5 arbor.

When the S3, S4, and S5 arbors have been thus aligned, turn the minute hand until warning occurs, and check that the hammer remains down: It should not lift until striking actually begins.

When striking is satisfactory, press the gathering pallet on firmly but carefully, and tighten down the S3B subplate.

Step 42. As the pressures on the escapements of French clocks are relatively weak, the pallets are often found to be in good condition compared to other clocks of similar vintage. Although smaller, the anchor of the French recoil escapement is similar to that of other European mantel clocks, and is set up in the same way. See **Escapement** ("Recoil—Solid Pallet—English, French, and German)" in Part IIII.

Step 43. Give each mainspring about two turns, and check that both trains will run.

Step 45. Oil the movement as indicated under **Oiling** in Part IIII. It will be noticeable that only the largest pivots are long enough to pass through the full thickness of the plates. The smaller, shorter pivots have oil sinks surrounding their pivot holes, which gives them a shorter bearing of reduced friction. Do not fill these oil sinks to capacity, however; use only a minimum of oil, as excess oil lying on an exposed surface will dry up before it can be useful to the pivot.

Fit the bell post and bell on the rear plate, and adjust the arm of the hammer, as required, to obtain a clean, sharp ring.

Step 47. Replace both hands and synchronize them with the striking. The hour hand may be moved to the appropriate hour, either forward or backwards, but this must be done with care, as breakage is always a possibility.

Stand the case on a level surface, and see that it cannot teeter. Reinstall the movement and level it. (A good way of doing this is to set up a spirit level in front of the dial and then rotate the movement until the nine o'clock–three o'clock axis lines up with the level.) Do not tighten the two strap screws too much or the case may distort.

Step 49. Hook the pendulum on, and set the clock in beat, taking the bell off its post if necessary. When adjusting the crutch arm, grip it firmly with the pliers just below the suspension spring and bend the lower end to left or right, keeping the pliers centered to avoid damage to the escape-gear teeth. See **Pendulums** in Part IIII. If the suspension spring is damaged, it is best to either replace it with a new one or rebuild the old one by renewing both sections of it.

Three ounces is a typical weight for the pendulum, and a small arc is to be expected.

CHAPTER

7

German Clocks

It is unlikely that there will ever be common agreement on just where the mechanical clock was invented, but German endeavors in horology go back far enough in history to deserve serious consideration in this regard. Perhaps the first timekeeping hardware to appear in Europe did so in several places simultaneously, give or take a few years.

The older clocks of Germany, like those of England, are by no means common, and com-

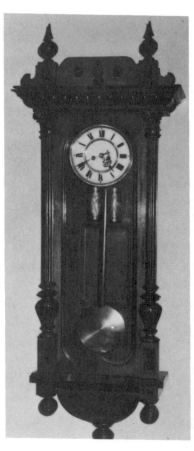

a.

b.

mand a high price. After several centuries of slow development, Germany eventually followed the lead of America and adopted mass production. The first such clocks are credited to members of the Junghans family, who imported American machinery to make the parts. In the years after 1862 production slowly climbed to rival that of America. When American clocks were first imported into England, English clockmakers received a mor-

Fig. 29. German Clocks
    a. Mantel
    b. Weight Vienna
    c. Mantel (after 1901)
    d. Box
    e. Spring Vienna

tal blow, and when a second invasion occurred, this time of German clocks, any hopes for the survival of the old methods surely died.

The earlier mass-produced models were largely copies of American clocks in both movement and case design, but it did not take many years for the Germans to develop both according to their own ideas. By 1890 large numbers were being produced to designs that bore no great resemblance to those of America. Because they were produced in quantity, it is these German clocks, rather than the earlier ones, which are largely available to today's clock addicts.

Most American clock cases are made of either solid wood or veneer. German clocks use a mixture of both, with large sections of fir and pine which are stained and varnished to blend in with the veneered sections. It is noticeable that the veneer on American clocks is often wrapped around a fairly tight curve, such as the frame of the door in a school clock, whereas the Germans generally veneered flat sections or curves of a larger radius, such as the bottoms of their wall clocks. German clock cases also make great use of turned columns of many sizes and shapes.

A wide variety of mantel cases were produced, of which that depicted in Fig. 29a is typical. Small hand-carved pieces were frequently glued on as decoration in both mantel and wall clocks, as shown in Fig. 29b and e. Simpler, more streamlined cases, such as those shown in Fig. 29c and d, came along with the art nouveau movement in about 1900. The movements, however, remained virtually unchanged.

8-DAY SPRING MOVEMENT

Movement: Fig. 30 Cases: Fig. 29a, d, e
Key sizes: 9, 10, 11

The movements of German mantel and wall clocks are basically the same, except for variations in geartrains and teeth counts to allow for different pendulum lengths. In view of this, only one repair text is necessary.

Like the American movement from which it largely developed, the mass-produced German movement has cutout brass plates, among other similarities. It is smaller, and the pivots are almost as fine as those of French clocks. The somewhat untidy free American mainspring evidently did not appeal to the systematic Germans, so they enclosed them in barrels. This step raised costs, but it also kept the spring cleaner, helped preserve its lubrication, and prevented any pressure on adjacent arbors.

The European preference for cut pinions prevailed here to some extent, but the lantern pinion was also widely used; sometimes both types are found together in the same movement. Even in the early days, when German movements resembled American ones, the no. 1 gears were usually made a little thicker—a useful improvement.

Although wire lifters were sometimes used, there was a marked preference for stamped-out, sheet-metal hooks mounted on brass collets. Even the click springs were generally of flat metal rather than wire.

The strip pallet was retained, but moved from outside the plates to between them; ready access was provided by a removable pendulum cock carrying the rear pivot of the pallet arbor. Oddly enough, the pallets were often left unhardened; however, the pressure upon them is somewhat less than on American pallets, due to the less powerful mainsprings. As regards mainspring power and the size and refinement of movements, the German product lies between those of America and France, and is a nice blend of both.

The following repair steps apply also to English mantel clocks, for when England belatedly began to mass-produce clocks, the resulting movements were much the same as German ones. There was, however, no quantity production of wall and longcase clocks to correspond with German ones; only the later art-nouveau mantel clocks seem to have been made in any worthwhile numbers in England.

Both rack-and-snail and count-wheel types were made; as the latter type is perhaps more common, it will be considered here. Fig. 30 shows a mantel movement, with mounting lugs on the front plate; however, a wall-clock movement would serve equally well as far as repairs are concerned.

Whichever type you are working on, first remove the pendulum and hands and take the movement out of its case in the usual way—by unscrewing its mounting screws. In mantel clocks it is often necessary to first take out the gong and its stand. The first few steps should therefore be performed in order of convenience.

Step 3 is as usual: Remove the pendulum suspension spring by withdrawing its pin; it can then be taken out of the hook at the top of the pendulum hanger, as the top of the pendulum shaft is called. "First off and last on" is a good general rule for suspension springs during an overhaul, as it prevents damage to this delicate part.

Carry out Step 6 if the movement is very

Fig. 30. 8-Day Spring Movement

oily and dirty; the oil must still be in the liquid state for a primary rinse to be effective.

Step 7. Check that the movement is complete and undamaged. The most usual damage is that caused by the breaking of a spring. Evidence that this has occurred is a spurted spring hook in a barrel, a bent no. 2 arbor, and a couple of broken teeth or trundles. (See **Tooth replacement—Going Barrel** and **Trundles**, if such damage is present.) Although the American mainspring is more powerful, it is not confined; thus when it breaks it has somewhere to go, and so causes somewhat less damage, making repairs easier to effect.

Step 8. Make a note of any pivot holes that need bushing; move the crutch up and down to check the pallet-arbor pivot holes.

Step 9. Actuate the strike train if possible by means of the count hook; if it locks satisfactorily, make a note of the at-rest positions of the warning pin, and so on.

Step 10. Scribe a small "S" on the strike-side barrel, barrel cover, and arbor; the time train is unambiguous and needs no marking.

Step 11. Let both mainsprings down fully.

Step 12. Unpin the hammer arm, and remove it from the hammer arbor. Take the hammer head off the hammer arm, and put it in dish no. 2, so that it evades the cleaning machine, which would erode its leather insert. If the lifter arbors are pinned or otherwise secured, remove their fastenings so that both arbors are free to come out of the plates. If the count wheel is outside the plates, pry it off using two screwdrivers (as suggested for removing a gathering pallet on page 73). Unscrew the pendulum cock and take out the pallet arbor. Remove the reduction-gear retaining disc, and take off the gear, the hour pipe, the two ratchet-gear retainers, and any other removable part from the front plate.

Unscrew the pillar nuts and set the movement on the assembly box. Pry the plates apart, and then take out the pin that holds the center spring in place on the minute arbor. Remove the spring and the T3 gear that it secures; the barrels can now be taken out, along with the other arbors.

Step 13. Check both barrel arbors for play. (See **Bushings—Barrels** in Part IIII if bushings are needed.) Take out both mainsprings along the lines indicated in **Mainsprings in Barrels** in Part IIII, taking care to scribe a small "S" on the soft outer end of the strike mainspring.

Continue with *Steps 15* and *16,* cleaning and inspection, respectively.

Step 18. Check that all gears and collets are tight on their arbors. If the gear is still on the S2 arbor, both it and the star gear on the S3 arbor must be reasonably firm, but not as tight as other collets, as both of these parts may need small adjustments later.

Step 19. Remove the fly from its arbor, and clean out its bearing surfaces. Make sure that it has adequate tension when replaced.

Step 20. Examine these pins and make sure that they are tight: The crutch-stop pins on the back plate, the warning pin on the S5 gear, the locking-hook stop pin (if present). The hammer tail and the stop spring, both of which are on the hammer arbor, must also be tight.

Step 21. Make sure that the pendulum cock is firmly riveted to its small subplate, and that the slot is vertical. Fit the suspension spring and adjust the jaws of the cock to give a light push fit that will still enable the bearing to swivel.

Step 23. See that the mounting feet and pillars are soundly riveted, and tighten them as necessary. Adjust the plates, as required, to ensure that they fit together easily.

Continue with *Steps 24* and *25,* which concern rust removal from arbors, lifters, etc., and treatment of rusty hands, respectively, as described in the ogee text.

Step 26. Examine both winding clicks and file their tips, if necessary, to ensure a good mesh with the teeth of the ratchet gears; check that all such teeth are sound, and that a minimum radius is present at the base of each tooth. The clicks may be of brass or steel, and while their rivets seldom work loose, check them just the same, for as in all clocks they are under considerable pressure.

Carry out *Steps 27* and *28,* pivoting and bushing, respectively, along the lines laid out in **Pivots** and **Bushing** in Part IIII.

Continue with *Steps 29* and *31,* which are, respectively, **Pegging out** and checking on the meshing of any arbors involved in Step 28.

Step 32. Check out the plates for flatness, and adjust as necessary. Warps are uncommon in these movements; as the plates are of rolled brass, their thickness will be constant. German movements are fairly tight, the arbor end-play being no more than about 0.020″. This is only slightly more than is usually present in French movements, but much less than is found in American ones.

Owing to differences in design, the usual order of Steps 33 to 38 has had to be altered. Bear in mind, however, that all steps retain their original functions—and numbering—throughout this book. Only the *order* of the next six steps has been changed here; the work done in each step remains the same. Perform them in the sequence given below.

Step 36. Clean out the hour pipe. As the hole

is too small for the usual cotton swab, use a suitably cut sliver of dry wood, making sure that no resin is present. Then slip the hour pipe on the minute arbor, and see that it is free to turn at any depth; there must be no sign of a bind.

Step 37. Inspect all parts again. Then lay the front plate on the assembly box for assembly.

Step 38. Insert the barrels first, and then set all the other arbors in place, but leave out the pallet arbor for the present.

Step 33. Fit the T3 (center) gear on the minute arbor, and follow it with the three-armed center spring. Make sure that the spring sits level and exerts adequate pressure when pushed down to the pin hole. Insert the pin. This pin is not tapered; it is held in place by a pressed-up rim in the center spring. If there is any doubt about its retention, use a piece of suitable wire and bend its ends to a tidy "S" shape; this will keep it from working its way out if the rim is inadequate, and is an improvement in any case.

Now complete *Step 38* by closing up the plates, working upwards from the barrels, and screwing the nuts down finger tight. Fit the lifters and all other appropriate parts on the front plate, and ensure that the click springs exert sufficient pressure.

Step 34. If the count wheel lies outside the plates, replace it lightly on its arbor (S2). Do not press it on tightly, as a later adjustment is likely in Step 41.

Step 35. Assemble the hammer, complete with head. See that the arm does not turn in its collet. As the collet is generally well designed, the arm is seldom found to be loose. Set the straight-wire hammer spring into position so that it assists the drop of the hammer; it must only assist weakly, or the strike train may be overloaded.

The hammer-stop spring mentioned in Step 20 is mounted on the hammer arbor; it is usually curved to provide an oblique contact with a stop pin or adjacent pillar. Check that this curve is such that while preventing unwanted hammer bounce it also has no tendency to jam.

The normal order of step numbers can now be resumed.

Step 39. Test out all arbors for end play and freedom of rotation. Check all gears for meshing (see Fig. 6).

Step 40. Fit the minute hand on, and rotate it a few times, checking meanwhile that the lifters operate correctly. Correct any malfunction, as necessary.

Step 41. Set up the strike. Wind the S1 arbor two clicks—*not* two turns—of the ratchet. This is done so that the mainspring will be less inclined to unhook itself from the arbor as the train is pushed round by hand.

In a typical German movement there are four arbors which must synchronize—namely, S2, S3, S4, and S5. However, since the count wheel and the star gear are moveable there is really only one critical meshing, S4 to S5, instead of the three that might otherwise be expected. As several variations in design are commonly found, a drawing corresponding to Fig. 17 would not be comprehensive enough to cover a sufficient number of movements. The various hooks perform the same functions, however, so reference to Fig. 17 may prove helpful.

Rotate the train until the cam-locking hook drops into the slot in the cam on the S4 arbor (some cams have a flat instead of a slot). In this position, the warning pin must lie against the pin-locking hook. Ensure that there is no residual power left in the mainspring, and then remesh the S5 pinion with the S4 gear, as necessary, until these arbors are synchronized. This can be achieved, even though the pin-locking hook cannot actually drop far enough to lock, because the count wheel is holding up the whole lifter via the count hook. This can now be corrected by adjusting the count wheel a slight amount forward or backwards until the count-hook blade can enter the nearest hour slot. Restore two clicks of spring tension as required.

The count-hook blade must land close to one side of the slot, as each slot is wide enough to accommodate both an hour strike and the following half-hour strike; hence the count wheel must be carefully set to allow this. If the count wheel is not located outside the plates, it is usually riveted onto the same collet as the S2 gear. In either case, it is simple to adjust. If it is external, turn it by hand; if it is internal, use a small screwdriver as a lever and apply it to the spokes of the count wheel and S2 gear. The leverage thus acquired is adequate over quite an angle. It is best to keep the lever largely parallel to the arbor, to avoid distorting the count wheel.

When the lifter drops satisfactorily, adjust the star gear on the S3 arbor, using similar leverage against the spokes of the S3 gear. As usual, when striking is over the hammer tail must drop off the tip of one of the star-gear teeth, leaving the hammer entirely free. It must not rise during the warning, but only when the train actually runs to perform the next strike sequence. Operate the train manually several times, readjusting the count wheel and star gear as required, until reliable striking is obtained over at least a full turn of the count wheel. Give the winding arbor two turns, and then run the strike train through several more sequences, using the minute hand as an actuator; this will check the operation of the whole mechanism, including the front plate components.

Step 42. Set up the escapement as described in **Escapement** (2) *Recoil—Strip Pallet—American (Internal) English and German.*

Step 43. Wind up the T1 arbor about two turns, and check that both trains run. Trains that run under low power may be judged to be in good condition. Sometimes, however, trains may be reluctant to run without a suspension spring and pendulum hanger.

Step 45. Oil the movement, as indicated under **Oiling** in Part IIII. German clock plates often have small oil sinks around most of their pivot holes. To be effective, oil sinks must be small. Larger ones do not really retain oil; they reduce friction because their presence shortens the pivot hole. Oil sinks are not intended to be filled to the brim with oil; use a minimum of oil at all times, whether a sink is present or not.

Step 47. Return the movement to its case, fit the hands on, and synchronize them to the striking. Make sure that the case is level, and that it cannot teeter across opposite corners.

Step 49. Install the pendulum, and set the clock in beat. Do not lever the crutch against its stop pins, as this may break a pivot or bend a tooth of the escape gear. Grip the crutch arm firmly just under the escape cock with a pair of long-nosed pliers, and then gently bend the lower end to the left or right. In most German movements, the crutch connects to its collet on the pallet arbor via a spring friction washer; this excellent idea greatly facilitates beat setting and prevents damage from excessive pressure.

VIENNA REGULATOR 8-DAY WEIGHT MOVEMENT

Movement: Fig. 31	Case: Fig. 29b
Crank size: 0	Weight: 3 lbs
Pendulum length: 21½"	Bob weight: 12 oz

As the name *Vienna regulator* implies, these clocks originated in Austria, in 1790 or thereabouts. The chief characteristics of these older, Austrian ones are a plain, high-quality case and a finely made movement with a deadbeat escapement. Often made in mahogany, they are somewhat rare, and command a high price today.

By the time the Vienna regulator had attained its first century of life, the Germans were making them in fair numbers. The high standard of the movement was unchanged, but the cases were mostly mass produced, became over-ornate, and were made of partially veneered fir, pine, and birch, in the manner that was popular prior to 1900. It is these clocks, still referred to as Vienna regulators, that can be found, after some seeking, by today's clock fanciers. (Note, however, that the type of case shown in Fig. 29b was also used to house spring-driven movements.)

Several types of Vienna movement were made, the one-weight T.O. and the two-weight T & S versions being the most common. (However, neither is exactly cheap today.) A lesser number of the three-weight, *grande sonnerie* type were also made; there appear to be even fewer of the miniature one- and two-weight Viennas, so presumably these were not made in quantity.

The Vienna is a relatively quiet, low-power device. One might say that it is a gentleman's clock—and his lady can easily wind it. Other people have other names for it, however, because unless everything is in excellent order, it has been known to cause great exasperation by stopping periodically. This is often due to dirty pallets. It seems to saunter through life, with its large, brass pendulum bob leisurely swinging at about eighty beats per minute over an arc which seems too small to survive.

When it comes to timekeeping, however, it is at the top of the class. A properly adjusted, single-weight Vienna is probably the most accurate of the world's more-or-less common domestic clocks. Although its rate may vary a little from summer to winter, due to the effect of changes in humidity on the wood of the pendulum shaft, a discrepancy as small as a minute in three or four months is not unusual.

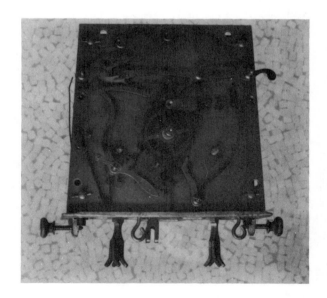

Fig. 31. Vienna Regulator 8-Day Weight Movement

Another factor that affects performance is thickening oil.

In view of its timekeeping accuracy, it is very strange that of all clocks the Vienna should have been singled out as the recipient of a bizarre aberration—the addition of a seconds bit (dial) which does not, in fact, indicate seconds at all. The seconds dial, placed below the figure *XII,* is marked for sixty seconds in the normal way, but the hand serving it completes a turn every forty-five seconds or so, and hence is useless as a seconds indicator. Without doubt many owners never even discover this irrational feature of the clock, but its presence on a fine clock made by some of the most methodical people on earth provokes some thought. It is fortunate that not all Viennas are thus afflicted.

The three-weight Vienna is somewhat uncommon, and its movement is relatively complex. The single-weight version is more easily found, and its movement is very much like that of an American banjo clock, save that it is more refined and has a deadbeat escapement with the pendulum at the back. The two-weight T & S Vienna seems to be the most numerous; its movement is shown in Fig. 31.

If Fig. 31 is compared with Fig. 26, a marked similarity will be noticed; the two-weight Vienna movement is indeed very much like a refined, scaled-down English longcase movement. It is a fairly tight movement, its tolerances and pivot sizes being similar to those of the French 8-day spring movement. As both trains run on a pull of only about 1½ pounds (3 pounds compounded once), friction losses must be held to a minimum if the clock is to keep running.

As both the cases and movements of Viennas are somewhat delicate, they do not take kindly to travelling, so a few precautions are strongly advised. Since the pendulum is not always easy to remove, it is best to secure it to the back of the case with masking tape. (Two vertical strips, one at each side of the bob, and a third, horizontal strip across the pendulum shaft just above the bob constitute a good arrangement.) Remove the masking tape at the earliest opportunity, since if it is left on for several weeks it will be difficult to remove and may mar the surfaces involved.

Inside the clock case, screw a small hook or slightly opened eyescrew into the base; use a fairly central location where it cannot interfere with the weights when fully descended. When removing the weights, keep some tension on their pulleys and then fasten them to the hook with an elastic band and a couple of paper clips or small pieces of wire. This simple step prevents the troublesome tangle of cable that otherwise always takes place around the winding drums when weights are taken off and cables are allowed to relax. At the end of the journey, it is best to eschew further winding and allow the clock to run down (if it will). Otherwise, if the cable has shifted on the drum in transit, immediate winding can still result in a tangle.

Access to the movement is good, it being only necessary to slide it out after unscrewing the two knurled knobs that secure its seat plate to the mounting bracket. Sometimes a rather less satisfactory method is used and four pillars have to be coaxed out of four keyholes in the backplate of the movement; these are sometimes troublesome to unfasten. In either case the pendulum remains behind, because it swings from the bracket instead of the movement.

Step 1. Unpin the hands and take them off. The so-called second hand, if present, is mounted on a pipe on the escape-gear arbor, and is simply eased gently off with a slight rotary action.

Step 2. Unpin the four dial pillars and remove the dial.

Step 6. Give the movement a primary rinse if it is excessively dirty and oily—provided that any oil present is sufficiently liquid to be removed. Remove the seat plate to clean it.

Step 7. Examine the movement for wear and

damage; because so little force is applied to the trains as the movement runs, they wear well and have a less-than-average need of bushing.

Step 8. Make a note of any pivot holes that need attention; because the plates are well finished, it is not a good idea to mark them. Unscrew the pendulum cock and take out the pallet arbor.

Step 9. Make a note of the positions of the warning pin and the other striking parts when the train is at rest. (See Step 9 of the English longcase text, page 73.)

Step 10. There is no need to mark any of the gears, for no two are alike. Unscrew and remove the rack spring, the snail's star-gear detent spring and the reduction-gear cock. Unpin and take off the remaining strike components from the front plate. Ease the gathering pallet off, as previously described in the English longcase text, page 73.

The components of both the Vienna and English longcase movements bear the same names and perform the same functions, even though their designs differ slightly. The Vienna, unlike the English longcase, has a separate arbor for the snail, with an attached star gear which is actuated by a pin on the cannon pinion and positioned by a detent arm and roller. This arrangement positions the snail more precisely than does the somewhat looser configuration used in the English longcase; however, both variations work well in practice.

Unpin and remove the hammer from its arbor, and put the head with its leather insert into dish no. 2, so that it does not go into the cleaning machine. Unscrew the two crutch-stop pins positioned close to the bottom edge of the back plate.

It is interesting to note that a serial number is usually stamped into both plates of these movements. Numbers such as 109560 are typical, but it is not safe to assume that this is any indication of the number of movements made, since we do not know the first number of the series. Frequently, there is not even a name or trademark to tell us who made the movement. Without such records, it is difficult to estimate just how many were made.

Step 12. Lay the movement on an assembly box, unpin the pillars, and gently pry off the front plate. Take out the arbors, and unscrew and remove the hammer spring. If screws are used to hold the pillars to the back plate, unscrew them and take out the pillars.

Scribe a small "S" on both the drum cover and spring collet of the strike drum, and then unscrew and remove these parts from both drums. Removing the drum cover releases the cable from the drum; unless the cable is in very good condition, it should be discarded.

The ends of some Vienna winding drums have two notches of differing size in them. The large notch will pass the knot which holds the cable; the small one will not. Both notches have their free ends covered by the drum cover, thus making them in effect D-shaped holes. In normal service, to renew a cable on drums such as these, the old, worn-out cable is clipped off short and pushed inside the drum, leaving both holes clear. The new cable is then knotted at one end, and the other end is fed into the large hole and out through the small one; the knot is then pulled into the drum through the large hole, where it is stopped by the small hole—all without taking the clock apart.

Unfortunately, the ends of certain other Vienna winding drums have only one hole in them, and to renew their cables the cover must be removed. As there is no access to the screw holding the cover in place, there is little choice but to take the whole clock apart—simply to renew a cable.

It is not a good idea to attempt to drill a second and larger hole in the drums while they are in place, as damage can easily result. A much better way is to drill two access holes, about 0.18″ in diameter, directly above the winding-arbor pivot holes in the front plate. (Always use a geared hand brace, never an electric one.) These holes must be marked carefully, to lie accurately over the heads of the cover-retaining screws. By means of these holes the screw can be slacked off enough to allow the cover to release the old cable knot and accept a new one; it is then tightened up again and the job is done. This arrangement is superior to the two-notch design, as it does not involve cutting a large notch in the drum, and also enables the easy removal of old cable remnants.

Although it is best to drill these two front-plate holes during an overhaul, when the plates are apart, it can also be done—very carefully—on an assembled movement, using a slip of sheet metal to protect the drum cover and the screwhead from the drill. Clamp the front plate in the vise, with appropriate protection, and *always* use a geared hand-brace to drill the holes. A drill press is excellent, even essential, for most clock jobs, but there is too much likelihood of damage for it to be used in this particular job. An assembled movement usually cannot be held in a drill vise, and to attempt to hold it in one hand would be buying trouble. Using a bench vise, as indicated above, also keeps the swarf out of the movement.

The use of the ubiquitous handheld electric drill must not even be considered for drilling holes in clock parts, as such machines are inaccurate and very destructive to the drill used in them. They should be used only for rough work with rough drills.

In addition to the usual winding ratchet gear and click, the T1 arbor usually carries a second such gear which engages with a separate click. In conjunction with a spring, this device keeps the pressure on the time train during winding, and so keeps the clock running. It is a common way of providing what is known as **maintaining power.**

Step 15. Put all the components of dish no. 1 through the **cleaning process.**

Steps 16 and *18.* Inspect the cleaned parts carefully, as advised in the ogee and English longcase texts. In examining the teeth of the rack, remember that the first tooth is intentionally short.

Step 19. Check the fly, and see that it is clean and correctly tensioned. If a groove has been worn in the spring and weakness results, replace it with a piece of old watch mainspring of the same thickness as the original.

Step 20. Make sure that these pins are strong and firm in their settings: the rack pin (on the rack itself), the rack-stop pin, the unlocking pins (on the reduction gear), the snail pin (on the cannon pinion), the warning pin (on the S4 gear), the hammer-lifting pins (on the S2 gear), the maintaining clock-stop pin, and the short bevelled pin on the rack tail. Check that the sloping pins (when present) on the drums are correctly placed, as in Step 20 of the English longcase text, page 74.

Step 21. Test the pendulum cock for firmness in the bracket. If necessary, rivet it tight, making sure that the slot is vertical. If the arms of the mounting bracket are loose, take the bracket out and hammer them tight again. See that the wood screws that secure the bracket to the case back still have a firm grip in the wood of the case. Adjust the jaws of the pendulum cock as required to suit the pendulum suspension spring.

Step 24. Remove any rust from steel components, as instructed in Step 24 of the English longcase text, page 74.

Step 25. Deal with the hands as in the ogee text. These hands in particular are somewhat fragile and easily damaged, so use considerable care in applying the fine emery cloth. See that the "seconds" hand is tight on its pipe. If riveting is necessary, take care not to distort its split pipe.

Step 26. Examine the clicks and ratchet gears for correct meshing. Since only light pressures are involved in the Vienna movement, these parts are not generally troublesome. The maintaining-power click leads a relatively easy life and seldom loses its shape.

Continue with *Steps 27, 28,* and *29,* **pivots, bushing,** and **pegging out,** respectively. Take care in Step 28 to preserve the original center positions; a small error that would not affect an American clock will stop both a French clock and a Vienna. If the pegwood should break, see Step 29 of the French 8-day spring text, page 88.

Step 31. Try out the meshing of any gears and pinions that may have been disturbed by any bushing done in Step 28. As the pivots are slender and fragile, use minimal pressure when closing up the plates.

Step 32. Check out both plates for warps and correct them as required. As the plates are quite substantial considering the size of the clock, warps are seldom found. The pillars are sturdy and relatively short; any force sufficient to bend them would surely have wreaked considerable damage on the rest of the clock.

Follow the French 8-day spring text, starting on page 89, for *Steps 33, 35, 36,* and *37.* These respectively deal with the cannon pinion, hammer, hour pipe, and a final inspection.

Step 38. Reassemble the T1 and S1 arbors. Adjust the spring collets to give a reasonable pressure, so that the drums turn easily for winding. However, the No. 1 gears must not be unduly loose. As the spring collets are usually held in place with a screw, rather than a pin, it is a simple matter to enlarge the screw hole, as needed, to arrive at the configuration advocated for the English longcase no. 1 arbors in Step 26, page 75.

If the pillars were removed from the back plate, screw them back into place, taking care that the two bottom pillars are the ones tapped for the mounting screws.

Now put the back plate on the assembly box, and screw the hammer spring in place. Set the minute arbor and the S2 arbor in place; then add the no. 1 arbors, followed by the others. Leave out the pallet arbor at present.

Working from the bottom of the movement upwards, close up the plates very carefully, taking care to avoid damaging the frail pivots of the upper arbors. Insert all four pillar pins lightly.

Screw the lifter tail onto the lifter, and then fit it and all the other front-plate components into place (see Fig. 31). Do not put the reduction gear in place as yet. Adjust the rack spring and detent spring as necessary until they exert just sufficient pressure for the reliable operation of the rack and star gear. Push the gathering pallet temporarily onto the projecting squared end of the S3 arbor, and then pin the hammer on its arbor.

Step 39. Test all the arbors for end play and all gears and pinions for backlash. See that no binds occur during rotation (see Step 39 of the English longcase text, page 78). Check that the lifter tail can move freely on its arbor, and that the roller of the detent rotates easily when the star gear is turned counterclockwise.

Step 41. Aside from certain small differences, the strike is set up in the same way as that of an English longcase movement; however, as the snail is on its own arbor rather than on the hour pipe, it is possible for it to desynchronize. However, this seldom happens; when it does, a weak detent spring is usually the culprit.

The snail collet also carries a twelve-toothed star gear, and this is actuated once an hour by the snail pin on the cannon pinion. After each impulse from the snail pin, the exact position of the star gear—and therefore of the snail—is determined by the lightly sprung roller detent positioned below the hour gear (see Fig. 31). The exact time at which the snail clicks over is not critical, except that it must be well clear of the hour warning-striking period, at which time the rack comes into contact with the snail.

The time taken to actuate the star gear is about ten minutes, and since the half-hour strike does not involve the snail, it is best to arrange that the changeover occurs at about the 30-minute mark so that it is completed by 35 minutes past the hour or thereabouts.

Fit a minute hand on the minute arbor and turn it until the star gear just clicks over. Reposition the hand until it indicates a time somewhere in the 20–40 minute region. Bring the hand up to 60 minutes (dead vertical), and then mesh the reduction gear with the cannon pinion and the hour gear so that the tail of the lifter just drops off the hour unlocking pin.

There are two unlocking pins; the hour unlocking pin lies farther out from the center of the pinion than the half-hour pin. The half-hour pin causes the rack hook to rise only far enough to clear the short first tooth of the rack, thus allowing the gathering pallet to turn one revolution only.

Applying finger pressure to the strike train, turn the minute hand round to cause striking to occur. Adjust the tail of the rack as required to ensure that the gathering pallet counts cleanly with no tendency to land on the tips of the rack teeth.

The rest of the strike-setting procedure is identical with that of the English longcase movement, so see Step 41 (page 103), while referring to Fig. 31 instead of Fig. 27. The same three arbors (S2, S3, and S4) must be synchronized, but greater care is required with the Vienna, as the pivots involved are very much finer.

When the gathering pallet is oriented correctly, pin it in place. (Provision is usually made for this on Vienna movements, but only occasionally on longcase movements.) A pin is a good way of securing the gathering pallet, because it does not require the use of a hammer, which would mean that most of the force would be borne by the end of the back pivot or the back end-bearing of the arbor. This would be acceptable on the large, robust longcase movement, but not on the Vienna. A smart tap on the latter movement is not advisable.

When the strike is satisfactory, tighten up all tapered pins.

Step 42. Set up the escapement according to **Escapement** ("Dead Beat—Solid Pallet—Vienna Type") in Part III.

Step 43. Check that both trains are free to run under finger pressure. It should be noted that not all deadbeat escapements will run under such conditions, even when in good order; it is therefore enough to check that forward and reverse pressure applied to the T1 gear produces a corresponding small twitch in the escape gear. If the escape gear is the slightest bit reluctant to move, it is likely that there is a bind somewhere in the train; there is no point in proceeding further until it is found.

Typically, binds are caused by a damaged tooth, dirt or a small fragment between the leaves of a pinion, incorrect meshing, a bent or scored pivot, lack of end play, and pivot holes that are too big, too small, or scored.

Step 44. Replace the original cables, or, preferably, fit new ones to the winding drums (See Vienna Step 12). Nylon monofilament fishing line, 0.025" thick, makes an inexpensive, excellent cable for a Vienna movement. It is pliable, fits well in the spiral groove, is not hygroscopic, and wears well under normal circumstances. Strong sunlight reduces its life, but that is not likely to be a problem. As in Step 44 of the English longcase movement (page 79), round off the edges of the exit holes in the drums, to avoid possible damage to the cables. Leave the cables overlong for the present, since they will be attached later, in Step 47.

Step 45. Oil the movement sparingly, according to the guidelines given under **Oiling.** Do not neglect to oil the rear of the reduction gear, the S3F pivot, the clicks, and the detent roller.

Step 46. Inspect both pulleys (see Step 46 of the English longcase text, page 80). Vienna pulleys have not been typically subjected to heavy weights, but, like other clock pulleys, have frequently been allowed to run without oil, which inevitably causes excessive wear. If no repairs are needed, oil the pulleys now.

Step 47. Screw the seat plate to the movement, making sure that it is right side up; do not, however, fully tighten the screws yet. Pin the dial in place and fit the hands on, making sure that they cannot foul each other in passing.

Step 48. Apply a little pressure to the S1 gear and check the accuracy of the strike. If striking occurs either side of the 60-minute mark, the hand collet must be adjusted accordingly. Two pairs of serrated pliers are required for this, one to grip the collet and the other to grip the center disc of the minute hand. This is not always easy to do, for the collet is sometimes chamfered. The pliers must be in good condition for this task, and applied with care. Force exerted on the hand itself, even close to the disc, often results in a broken hand. The motion required is small. When this adjustment is accomplished, do not rivet the collet any tighter than is necessary for it to reset the hands in the normal manner, because someone else may need to reset it at a future date.

Step 49. Hang the case up and fit the pendulum to the bracket. Check that the gong is firm, and that its coils are free and do not foul anything. Push the pendulum gently—a 2″ total swing is quite enough—and watch it long enough to see that it is going to behave itself. (See **Pendulums** in Part IIII.) Step 49 of the English longcase movement (page 82) also contains some relevant information.

When the pendulum is performing in a satisfactory manner, slide the movement into the bracket, taking care that the crutch pin enters its slot in the pendulum shaft. Close the case door and check the position of the dial. If it is not centered, open the door, slacken the seat-plate screws, and center the movement. Repeat as necessary, and tighten the screws on completion.

With the door closed, make sure that the glass does not touch the end of the minute arbor. This is another small fault that will stop a Vienna. Naturally, the end of the minute hand must not touch the glass, either. Insert the two knurled screws which secure the seat plate, but tighten them only when the correct forward-backward position has been reached and the hammer arm has been adjusted.

This last procedure is done by operating the hammer gently by hand and adjusting the arm until a clean buzz-free tone is obtained from the gong or rod. It is sometimes best to remove the movement each time for this, as clearances behind the movement are not always generous.

Give about half a turn to the winding arbors, so that their cables hang from the correct side of their drums. Fit the pulleys onto the cables so that the decorative side of their yokes (if present) faces forward. Stand both weights on the bottom of the case and hook the pulleys to them. Run the free ends of the cables up to the movement, and make a loop in them which will fit neatly over the hooks provided. When the clock is fully run down, no more than about half a turn of cable should be left on the drums. If too much is left on, the final turns will overlap, which may cause a tangle.

Now set the case vertically on a wall, using a spirit level. Reposition the scale, if necessary, so that it centers just below the end of the pendulum-rating screw. Check that the crutch pin does not exert a backwards pressure on the pendulum; if it does, bend the top of the crutch arm closer to the movement.

With a forefinger on each side of the bob, move it about 1″ to one side and then allow it to swing. The use of two fingers keeps the bob parallel to the case back and does not impart an initial twist, as a single finger might. Bring the clock into beat by adjusting the small, knurled knob at the end of the crutch arm. Often there are two such knobs, in which case it is best to use both hands for additional steadiness and ease of turning.

The pendulum arc should settle down to about 1.5″ or 1.6″, with a slight increase after the clock has run for a few days.

III

RESTORING DIALS AND CASES

CHAPTER

8

Dials

There are many types of clock dial, and their restoration involves a variety of methods. The oldest dials likely to be encountered are the square dials of the early English longcase clocks. These are of beautifully engraved brass, and the slightly later break-arch type of ca. 1750–70 often had its second bits, date bits, and chapter ring silvered. Where necessary, re-silvering is feasible, but it is important to realize that most of these old brass dials were not silvered originally.

To be restored, a dial must be taken apart, and each piece cleaned, polished, and lacquered separately. It is best to use a cleaning machine for the spandrels and bits (the seconds and date dials).

After 1770 the white-painted steel dial replaced the brass one; and because white dials were made in considerable numbers, they are more common today. Also to be found are painted wooden dials, such as those of American clocks of about 1800 vintage. Later, painted zinc dials took over from the wooden ones, followed in turn by printed paper dials, doubtless as a cost-reduction measure.

German clocks of about 1880 often had porcelain dials, although a few were of chased or pressed brass. After 1900 the plainer box clock came along with a simple dial of silvered brass. When the thin silvering wears off these dials, resilvering is probably the best course of action, but another possibility is producing a spun-brass finish with fine emery paper (see page 113).

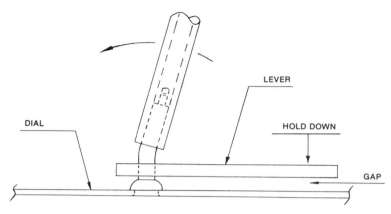

Fig. 32. Straightening a Dial Pillar

Longcase White-Dial Repairs

Painted clock dials, while indicating the passing of the hours, often show the ravages of the years. Due to many decades of polishing and dusting, the edges of numerals and other details on a longcase clock dial slowly wear away. The maker's name often disappears, and with it a good chance of accurately dating the clock. Often when no name is evident to a casual glance, it can be picked out as a faint, slightly raised, white-on-white series of letters shinier than the rest of the dial. Although no black paint remains, it once protected the underlying paint, leaving an outline which can be seen when light is shone on it at certain angles. It always pays to search for this, both before and after the dial has been cleaned.

Remove dirt from the dial by rubbing it lightly with a fine household-cleansing powder, used wet. The central area of the dial and the break arch are the best places to look for the maker's name.

It is almost always preferable to restore an old longcase clock dial, rather than strip it and repaint it. Background damage must be dealt with first. This is commonly found in the areas of the dial pillars, as any applied stress will loosen the riveting and break out the brittle dial paint.

A bent pillar can be straightened with a piece of ¾" × ¼" mild-steel bar; drill a close-fitting hole in the bar, near one end, and use it as a lever against a suitable pipe or tube, as shown in Fig. 32. Never attempt to straighten a bent pillar without supporting it, as it will usually pull right out. When the pillars are straight, rivet them tight again, using a well-fitting hole in an anvil. Check that the pillars fit into the front plate correctly, and adjust them with the lever and pipe as necessary. The hand arbors and winding squares must locate centrally in their dial holes.

Riveting usually knocks out a little more loose paint from the dial, and this must be replaced before any art work can be done. The two problems here are matching the color and levelling the riveted area with the surrounding surface. Color matching is a matter of trial and error. In most cases, a very small drop of black and a tiny amount of green added to the basic cream or light yellow will effect a good match. Matching is complicated by the fact that paint changes color slightly as it dries, and again when spray lacquered; allowance must be made for this.

It takes several coats of paint to build up the required thickness. When the new paint is level with the old, smooth out the last coat with a wet fingertip. This will obscure the outline of the repaired area. It is feasible to use fairly thick coatings initially, but each one, including the final one, must be rubbed down with fine wet-or-dry emery paper and water. To avoid runs, always lay the dial flat to dry it; an overnight stay in a warm place, such as on a hot air duct, will ensure good drying.

Oil-based hardware-store paints or enamels mix well and dry in a reasonable time. Artist's oil colors may seem a natural choice, but they take much too long to dry and harden. Even with the recommended paints several days are required to restore the background, so that repaired areas are satisfactorily hidden.

Touch up the colored areas of the dial where necessary, taking care to tone down the shades to match the original colors. Avoid bright colors, as they stand out as an obvious repair. However, if considerable repainting and reconstruction of the art work in the break arch and corner scenes is necessary, a knowledge of drawing and painting is required.

Provided you have a strong desire to do so, there is generally no good reason why you should not learn to draw and then to paint. Always ignore anyone who downgrades your abilities or potential; just go ahead and study what you want to learn. Perspective is the key to good drawing (see under "Recommended Reading" in the back of this book), and its importance cannot be overemphasized. Composition, light and shade, and so on, must also be studied. Only when you can draw well should you attempt to paint.

No matter how good the finished product may be in terms of concept, accuracy of drawing, definition, and execution, most painting is still a mosaic of small colored areas, cleverly shaped and positioned, on a flat surface. We need to be reminded of this, for our eyes have been so long trained to see pictures as representations of scenes that they no longer see what is really there—a mosaic. Study the details of these small colored areas on a sharp, well-executed picture, and see how they blend to give an impression of depth and distance and so create a three-dimensional illusion on a two-dimensional surface.

In restoring the numerals on the dial, note that much of the art work was evidently done in India ink or its early equivalent, and this must now be replaced. To do this it is necessary to determine the center of the dial. Cut a piece of flat lumber as big as possible while still fitting inside the dial pillars, and bond it to the back of the dial with two-sided masking tape. The thickness of the wood must be greater than the length of the dial pillars, so that they do not touch the bench.

Using a beam compass and any still-extant reference circle, such as the minute ring, determine the center of the dial by trial and error. Set one leg of the compass very lightly into the wood showing through the center hole of the dial, and adjust and readjust until a radius and a center are found which satisfy the reference points all around the dial. Deepen (very slightly) the center so found for future use.

Examine the dial carefully from various angles for traces of once-present black circles, etc., which, like the maker's name, many be present only as a fine, slightly raised shiny line inside or outside the minute ring. Such lines may be missing from the concave inside edge of the corner decorations, spandrels, for example. In some instances these may be gold or silver in color.

Keep the dial free of grease and wax which would make the ink bead and skip, and always use a piece of paper so that your hand does not touch the dial. Give all the areas to be inked a light prior rub with a soft eraser to ensure good adhesion.

Replace any missing minute marks, using a radially held rod or raised straightedge as guide; nonradial marks will be readily apparent. With the beam compass at the proper center and radius, draw in the missing circles and segments of circles, using a good grade of waterproof drawing ink. Avoid making the lines too thick, lest they stand out too much. When inking in the serifs of the Roman numerals, work from the serif ends towards the middle. Any excess of ink left behind when the pen is lifted from the surface will then tend to merge with the thick stroke of the numeral. Both Roman and Arabic numerals deserve more than a passing glance if they are to be drawn well.

Fig. 33 shows a typical Roman numeral of average width, where the *V* is about as wide as a minute space. (The *X* will be the same width.) The serifs shown are traditional, and should be copied; however, they do vary sometimes according to the taste of individual artists. Continuous serifs, for example, look well also, though they are only rarely found: The curves RS and QP in Fig. 33 would then be continuous. (The line QP must not overshoot P.)

There is a surprising amount of detail to note on a Roman dial. Note that the *I*'s do not have radial sides; they are all parallel to the radial hour line NO (Fig. 33). Remember this when drawing all these edges in anew (they will be mostly worn away, like other ink-work). This is best done with a ½"-diameter, round plastic rod, or round ruler, and a steel pen. Do not attempt to lay out these lines by measuring them; do it by eye, with the assistance of a few pencilled guidelines which can be erased later. Examine a sharp, clear Roman dial, and practise laying out dials to match it; note that each numeral is centered on an hour marker.

The *X* always lies in a rectangular frame; it does not taper towards the center of the dial. Two *I*'s occupy rather less than the space of a minute, and each is wider than the gap between them. In Fig. 33, the centerline of the *V*, line LM, is parallel to NO; however, the tip of the *V*, point P, is sometimes slightly off the line, as shown, in order to reduce the visual isolation of the *V*. Alternatively, extending the serif to the tip of a truly symmetrical *V* does the same thing.

The diamonds, triangles, blocks, or dots which mark the hours give an indication of the clock's age, for these and other details changed considerably over the years. For example, fewer minutes were numbered as dials progressed (originally, all were); finally, minute numbers disappeared. Dials with heavier-than-average Roman numerals tend to look clumsy, but the narrower, slimline type—which can be found on both wall and mantel clocks—is attractive, and gives the dial a refined look. The "height" of the numerals is unchanged in this type, but all thicknesses are

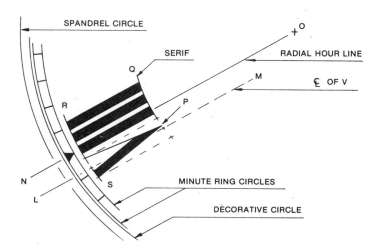

Fig. 33. Dial Detail

roughly halved. This slimmer style is rarely found in longcase clocks.

Arabic numerals probably look best when upright on a clock dial—that is, as normally written—but thus can never achieve the neat, radial appearance of the more popular Roman numerals. (It is just as well, however, that Roman numerals are largely confined to clock dials!)

Treat the seconds and date dials the same as the main dial, ruling in the seconds marks and so on, and renewing the Arabic figures by means of pen and ink or a little paint on a tiny brush, whichever you prefer. Do the same with the maker's name; however, this requires great care if it is to look well, for such lettering is not easy to do. If the name is surrounded by engrossing, practice long and hard elsewhere before attempting to fill in the gaps. Engrossing is rather like engraving in that it requires the swift, sure touch of a master who has had years of continual practice. Like good handwriting, it is a rare accomplishment, and since relatively few clock dials require it, very few people learn to do it.

When all the ink-work is satisfactory, touch up the black paint of the numerals, hour markers, etc. Give the dial about a day to dry thoroughly, and then erase all pencil guidelines. Remove any small outcrops of paint with a sharp knife, so that all edges are sharp. When all art work is satisfactory, spray the whole dial with a durable, clear lacquer, such as the acrylic final varnish that is used as a finish for oil and acrylic paintings. This will even out the differences between old and new paint, and also protect the dial and slow its aging process.

If it is necessary to repaint the whole dial, the bare metal should be treated with rust preventive and given at least five coats of cream or off-white oil-based paint. Each coat must be allowed a day to dry in a warm place, and then rubbed down flat with fine wet-or-dry emery paper and water. The final coat should be finished in the same way, until it is flat, matte and free of discoloration and blemish.

Now select the center, and with a beam compass pencil in the corner areas, and paint them in. Acrylic paints may be used for this, provided the dial background is prepared as above and allowed to dry thoroughly beforehand. Since the type of scene depicted in the corner art work changed every few decades, take care to date your clock and paint in the correct type of art to match its age (see "Recommended Reading" in the back of this book). For instance, gold and silver paints were only used extensively in the break arch when the English longcase clock was waning, chiefly from about 1850 to 1870, a period in which it became somewhat ill-proportioned and over-embellished.

Wooden Dials

Wooden dials are repainted in roughly the same manner as steel dials. Arabic numerals should be considered as each lying inside its own circle (see Fig. 35 below and its accompanying explanation). For removing warps, see Chapter 9, pp. 116 ff.

Zinc Dials: Stripping and Repainting

On both sides of the Atlantic ocean, zinc was widely used for clock dials, and because these dials oxidize readily they now are sometimes in very poor condition. The paint on these dials is usually blistered, cracked, or powdery; in most cases it is a waste of time to attempt to retouch them, as the paint has lost its adhesive properties and will simply continue to fall off.

In a few cases—such as ogees, certain German wall clocks, and Seth Thomas no. 2 regulators—reproduction aluminum dials are available. These are generally good copies of the originals, and they solve the problem at a reasonable price; however, they do not suit every clock or every owner.

The more valuable the clock, the more de-

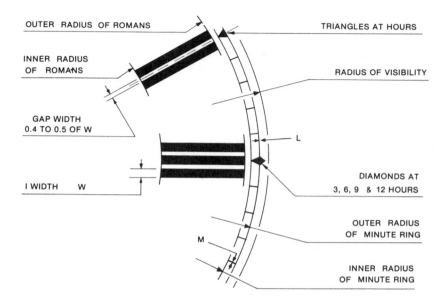

Fig. 34. Dial Data

sirable it is to repaint the dial in its original format. The dial of an Anglo-American wall clock is a typical case in point. Its characteristics must be noted from such markings as remain visible. One way of doing this is to make a tracing; another is to use measurements. Tracings are almost mandatory for lettering and logos, but for the usual, uncomplicated dial, measurements are much faster and entirely satisfactory. Always record the dial data; the object is to make a copy of an old dial, not a new dial to a new design.

Fig. 34 shows most of the required parameters. Match the line thickness L, and note the minute markers M, which are often somewhat thicker. The radius of visibility is the radius of the circle of visibility—that is, the maximum visible dial, which is usually the inner radius of the bezel ring. There is no need to draw this circle, but it must be noted in order to verify the size of the minute ring. In an Anglo-American wall clock, the outer radius of the minute ring (r_o) is usually close to 5½″, thus making an 11″ dial.

A convenient format for recording data is as follows:

Radius of visibility = _____
Minute ring:
  Outer radius r_o = _____
  Inner radius r_1 = _____
  Minute markers: (Lines, dots, etc.? Trace samples from dial.)
  Hour markers: (Blocks, dots, triangles, etc.? Trace samples from dial.)
  Three, six, nine & twelve
    o'clock markers: (Blocks, diamonds, dots, triangles, etc.? Trace samples from dial.)
Roman numerals:
  Outer radius R_o = _____
  Inner radius R_1 = _____
  Width of I's: (Four I's occupy 1.7 minutes?)
  Width of V & X: (One minute? Slightly less?)
Line thickness sample:
  r_1 & r_o:
  Other circles, etc.:

When the data is on record, make tiny indents on the dial with a scriber at the three, six, nine and twelve o'clock hour markers, and remove the dial grommets from the winding holes. This is done by prying up the rear lip with a triangular section-scraper or a putty knife; the grommets can be used again, if removed carefully.

Put the dial on a pad of newspapers, and scrape off the remaining paint with the end of a putty knife. Clear small areas at a time. When no paint remains, gently tap out any dents.

Go over the dial with a felt-covered wooden block and some fine, open-grain sandpaper; this does not clog up as easily as emery cloth. The idea is to remove most of the dark-grey oxide and provide a key for the paint. It is a good idea to etch the surface a little with rust preventive or some other weak acid, since paint does not adhere well to very smooth surfaces. Take off any grease, paint, or loose dirt from the rear of the dial, but be careful not to remove the names of former owners or repairmen. Such names are part of the clock's history, and the dates beside them help tell how old it is. Wipe the front of the dial with acetone or naphtha, to ensure that it is free of grease or residues.

The dial must now be given either several coats of spray paint or four or five coats of brushed-on oil-based paint of a suitable cream or ivory color. Each coat must be allowed to dry throughly before being lightly rubbed down with fine wet-or-dry emery paper and water. Do not use a high-gloss paint, as a matte final coat is required.

Fasten the dial in the normal position to a large piece of cardboard, using three or four small pieces of masking tape around its extreme edge. Determine the center by means of a straightedge and the four tiny indents at the three, six, nine and twelve o'clock hour markers. Check that this center matches that of the circle of visibility and is reasonably close to the physical center of the dial. A very small shift is sometimes required to achieve optimum center position.

Set up the beam compass from your dial data and ink in the minute ring. Check that the four indents are equidistant from each other, and in accord with positions of the winding holes. Pencil in the inner and outer circles on which lie the serifs of the Roman numerals. Pencil in the exact positions of the four indented hour markers; they may not coincide exactly.

To lay out the eight remaining hour markers and the minute markers, it is handy to have a transparent template. This typically shows a series of seventeen concentric circles with radii in ¼" increments from 2" to 6" long. These circles are crossed by sixty equispaced radial lines drawn in from a larger circle of 7" or 8" radius. However, only the hours are continuous lines; each minute is marked on each circle. Every fourth circle may be color coded for easy recognition of whole-inch radii. A 2"-radius circle might well be red, the 3" orange, 4" yellow, 5" green, and 6" blue, for example. Once carefully executed, such a template facilitates the quick layout of any dial of up to 12" diameter. The template is simply taped accurately to the dial and the minutes then pricked through along the minute ring. Once inked over, the tiny indents do not show.

Once made, the template provides a fast, easy, and accurate layout method. However, it is quite feasible to lay out the dial itself. Use a large protractor to pencil in the other eight hour marks. Check that all twelve are equally spaced by stepping around them with a pair of compasses or dividers. By trial and error, set the dividers to a fifth of the arc between two hour markers, using the outer circle of the minute ring, and "step out" the minutes with pencil dots. With a little practice, this can be done by eye; small corrections are then made at the inking stage.

Ink in the minutes. If they are dots, try to make them of equal size; if lines, of equal thickness, and take care to rule each one accurately from the center—errors will show. It helps to stick a sewing pin or needle in the center hole, and rest the straightedge against it.

Leave the hour markers as lines for the present; this will facilitate centering the numerals.

Lightly pencil in the arcs for serifs of the Roman numerals; continuous circles are acceptable, but more erasing will then be required later. Taking care not to smudge the ink-work, pencil in the Roman numerals, using the previous text and Figs. 33 and 34 as a guide. Note that, going clockwise around the minute ring, the heavy line of the *V*'s and *X*'s comes first, and both these figures look better if a trifle too narrow rather than a trifle too wide. It is helpful to rotate the dial when laying out the numerals, but it is a good idea to mark its exact position on the cardboard before removing the masking tape, so it can be replaced accurately later.

Now ink in the numerals, correcting the pencilled lines slightly, as necessary. Center the dial again on the cardboard to ink the serifs; do not let them extend too far.

Pencil in the hour markers, taking care that they are centered and of equal size, and then ink over the lines, balancing their outlines as required. Take a good-quality artist's brush of no. 1 size or less, with a good point on it, and paint in the numerals and the hour markers, keeping the brush fairly upright. If you accidentally overstep the inked lines, the resulting bulges can usually be put back with the other end of the brush. They can also be scraped off later with a small, sharp knife, when the paint is dry or nearly so.

Allow the paint to dry thoroughly, and then erase any pencil marks with a soft eraser; it will not disturb the ink, and it does not matter if it dulls the paint here and there. Tidy up any ragged edges, using pen and ink and the knife.

Brighten up the brass dial grommets with fine emery cloth, and fit them into the winding holes. To round them out in the holes, insert a pair of closed long-nosed pliers into them from the front, and spread the pliers apart slightly. Now set the front of the grommet on an anvil, over a hole almost equal in size to that in the grommet; keep the dial from contacting the anvil or scratches will result. Now spread the back of the grommet by means of a steel cone such as an oversize center punch and then gently tap or push it down, evenly all around, with a small ball hammer. Very little force is required for this, as it does not take much to flatten the front of the grommet and cause it to bite into the new dial paint.

Give the dial a final check for scratches and aberrations of ink or paint, and then spray it with a good clear lacquer of the type previously mentioned (page 110).

Smaller dials, such as those of kitchen clocks, are done in a similar manner. Slightly greater accuracy is required here, because errors stand out more on a small dial. When inking circles on convex dials, angle the pen so that it is always normal to the surface, as both sides of the pen point must make contact.

Paper Dials

When fitting a new paper dial, mark the six and twelve o'clock positions on the dial pan with a pencil, for ease in locating the new dial. Then strip off all the old dial paint, and prepare the surface so that it is flat and free from dents and humps; otherwise these will show through the new dial. For the same reason, take care to spread the glue evenly. A rubbery glue such as contact cement is preferable to a hard-setting glue, as the latter tends to lose adhesion and crack off the metal.

Silvered Dials: Spun-Brass Effect

A worn silvered dial, such as is sometimes found on a German box clock, can be refinished to give an attractive spun-brass effect. First make a note of the dial data, as always. Then remove the dial from its bezel, and take out the winding grommets. Fasten the dial down with two-sided tape to a piece of fiberboard (about ½" thick) and mark the center with a scriber.

Now cut a piece of smooth softwood (about ⅝″ × 1″) to a length equal to the outside diameter of the dial. Locate the center of the 1″-wide surface of the board, and drive a thin 1″ nail squarely through it so that it protrudes on the far side. Take a piece of fine emery cloth 3″ or 4″ wide and a little longer than the piece of wood and push the sharp end of the nail through its center, so that the emery cloth can be folded up around the wood like a sanding block.

Push or tap the nail into the center hole in the dial, so that the emery cloth rests on the dial. Now rotate the wood, under mild pressure, to rub the emery cloth on the dial surface. What is left of the silver will soon wear away, leaving a nice spun-brass finish. When this finish is uniform, carefully blow off any particles of dust, brass, or emery, wipe the surface with a clean, soft cloth, and spray it immediately with clear lacquer. When thoroughly dry, this antitarnish coating will also provide a good surface for the art work.

Quite rarely, a dial will be found with silver plating several thousandths of an inch thick, in which case a spun-silver finish is both feasible and desirable. Use the same technique on the silver.

In doing the art work, note that pencil marks do not show up well on clear lacquer. Mistakes are not easy to correct or hide, so painstaking work is essential. Roman numerals are laid out along the lines previously described.

Arabic Numerals

Arabic numerals, also to be found on these and other clock dials, pose a problem of uniformity when drawn by hand. There are five *1*'s and two *2*'s for example, on the clock dial, and any differences between them are quite noticeable. Dry transfers are easily applied and look well, but choice is somewhat limited, for most of them are quite unsuitable. One possible typeface for Arabic numerals on 6″ and 7″ dials is Times Bold, in 72-point size—although 68-point compact numerals group double figures better.

Like Roman numerals, Arabics must be positioned accurately. To do this, it is best to consider each of them as being centered in its own circle. These circles must be the same size and equispaced around the dial. A template for laying out the numerals is shown in Fig. 35.

The template is made of thin, clear plastic—say, 0.030″ thick, and has a lightly scribed centerline on it. On the centerline are the guide-hole for the numerals and several spaced pinholes for pinning the template to the center

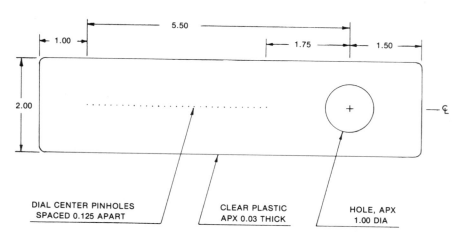

Fig. 35. Arabic Numeral Template

of the dial. When thus pinned at the required radius the template should be taped down with its centerline on each hour marker in turn. The numerals are then centered one by one in the guide hole and transferred to the dial underneath. Dry transfer sheets usually have dotted guidelines on them, and these can be used to keep the numerals level. This is done by lining up any convenient guideline across the dial, with any level pair of corresponding minute markers—say, 22 and 38 minutes, or 9 and 51.

Burnish the transfers down thoroughly to get the whole figure; any cracks or missing pieces can be filled in with a pen and India ink. When all is satisfactory, spray the whole dial again with the same lacquer as before. Two light coats should be used; a single heavy coating may either run or cause a numeral to shrivel slightly. (However, any such crinkles can usually be flattened out before they dry again.)

It is usually better to repair an old dial rather than completely repaint it, but when an old dial has become an unreadable eyesore it downgrades the whole clock. Despite the outcries of the originality fanatics, the dial should be renewed. It is perhaps the only part of a clock that is looked at many times a day by everybody. It takes a clock fancier to examine the rest of a clock; others are content with the dial.

CHAPTER

9

Cases

Repairs to wooden cases vary, from replacing a small piece of veneer to reassembling the whole case and completely refinishing it. As with dials and movements, it is best to preserve as much of the original as possible. The general procedure is to take out any warps, make the case structurally sound, and then deal with surface items such as moulding, trim, veneer, and finish.

As the same woodworking principles are used in the construction of all clock cases, it would be pointless to select more than three or four to serve as examples of typical repair problems; however, it must be clearly understood that the information given here applies to other wooden cases as well.

Longcases

The longcase is an obvious choice for consideration. As it is an older type than most clocks, it often exhibits almost all of the faults found in clock cases at large. Examination of the inside of a longcase will generally show a somewhat rough construction which is not always as strong as it should be. When a case has seen the passing of two centuries or more, it is understandably not as robust as in its younger days. It is therefore advisable to treat all such cases gently, and to avoid any shock treatment.

Owing to the presence of old and friable glue, pieces of the case may fall off quite readily. I once saw this happen as a case was being loaded into a truck, and the dealer cheerfully commented, "That figures. They self-destruct in about three weeks!"

Look the case over carefully to see what is loose. The waist is sometimes loose on the base section, and this must be corrected. Damage due to wood beetles is likely in English woodwork, and it is almost always worse than the surface holes appear to indicate. All such weakened internal pieces must be removed and replaced with sound wood. Remove entirely any loose pieces, inside or outside, for it is not feasible to simply pour glue down the crack and use clamps to make a joint, as this does not make a sound job. Each piece, and its mating surface on the case, must be thoroughly scraped or filed clear of all old glue, straightened to a good fit, and then reglued with brown glue again. Do not use white glue, since it will not bond properly with brown-glue residues; however, it is excellent when used with new wood.

Warps in Longcases

Before you can obtain a properly rigid case, it is usually necessary to remove several warps. This can sometimes be done by closing the

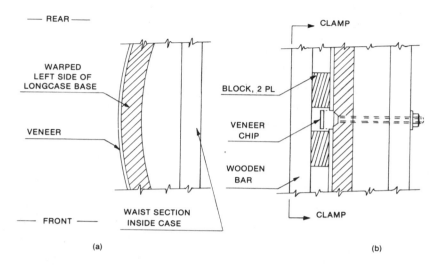

Fig. 36. Correcting a Warp, Using a Screw

crack with clamps, and then inserting wood screws to hold the pieces permanently in the clamped position. This method is far better than attempting to use the screws themselves to close the gap; this usually fails because the screw heads, under pressure as they turn, simply dig into the wood instead of drawing the two pieces together as intended.

Because warps occur frequently and their correction is essential, if not always easy, they deserve considerable attention. Although occurring in other locations, warps in a longcase are typically found in the following places: panels and mouldings of the base, waist door, waist panel above the door, mouldings at the top and bottom of the hood, hood door, and swan necks or fretwork on top of the hood. If the whole case is twisted—and it sometimes is—it is perhaps just as well to leave it alone. It may have been out of plumb even when made, and to correct it would entail taking the whole case to pieces, truing each part, and then rebuilding everything. In most instances, this is not justifiable, in view of the time and expense involved.

A warp is sometimes found at the top of a base side-panel (Fig. 36a). Since access that would allow cleaning off the old glue is almost nonexistent, it is better to use the screw technique. It is neatest to use a wood screw from inside the case, but the veneered wood of the side panel is not always thick enough to hold the threads. In this case it is best to use a screw and nut. This method, although unusual in cabinetmaking, also has the advantage of permitting later adjustment.

Carefully chisel out a single small piece of veneer for each screw (one or two is usually enough). Then pull the panel into its correct position by applying clamps across the width of the base. This does not generally involve much force, as the wood grain is vertical.

Drill and countersink the necessary screw holes. When the screw is in, tighten the nut from inside the case. At the top of a base side-panel the side of the waist extends down internally, and so can provide an anchor for the screw (Fig. 36b); in most other locations no such anchor exists, and screws cannot be used in this manner.

Glue the veneer chips back into place and clamp them firmly to keep them flat. Only two clamps are necessary throughout, one at the front of the case and the other at the back; between them they clamp two bars of wood that lie across the side panels from front to rear. A small block or two is also required to localize the pressure (Fig. 36,b). Two people can set up this clamping arrangement much easier than one.

Note that it is not much use trying to reglue such a joint unless the whole side of the base is first taken off for cleaning and renewal of *all* joints involved.

The use of steam or very hot water is an old and well-tried method of removing (or installing) warps in solid wood. Impressed patterns, such as those on kitchen-clock tops, were also made this way. When veneer is present, however, hot water must be kept off it; nor should it be allowed to soak through to veneer from the rear.

In many cases, warp removal involves making saw cuts in the back of the wood to reduce its strength. The wood is then straightened and clamped in position until the glue takes over and holds it in place. Hot water poured into the saw cuts and allowed to soak in for about fifteen minutes, will greatly reduce the likelihood of fracture when the panel or strip is straightened out. This is done on the bench by means of blocks and hand pressure, clamps being used when necessary. As the process is really a form of controlled smashing, a few protests from the wood are usually heard. However, surface damage is not usually severe, and can always be repaired should it occur. Fear not.

Fig. 37a shows a section of warped bar or moulding with saw cuts in it to allow it to lie flat as in Fig. 37b, where it is shown glued back in place. The saw cuts must not reach the veneer, but should stop about 0.16″ from it.

It is best to make a cut at the point of maximum curvature, and then several equispaced cuts over the rest of the curve. Adjust each cut as necessary to straighten the wood. The cuts of Fig. 37 close up somewhat on straightening, but the wood will be structurally weak until it is glued in place again.

Fig. 38a shows a wood section that is curved in the opposite way. However, the same general ideas prevail as regards depth and location of cuts, and so on. Since these cuts open up on straightening, they can be filled in with strips of wood which taper in cross section. These wedge strips should be shaped to fit each saw cut, and then glued in place; they are then planed and sanded flat (Fig. 38b). The resulting bar, panel, or wooden dial will be almost as strong as it was originally.

The warps shown in Figs. 37 and 38 must be thoroughly subdued—even slightly overcorrected—before the piece is replaced in the clock case, as glue is not strong enough to hold the slowly increasing pull of a determined warp. "Anti-warp" saw cuts must always lie parallel to the axis of the warp, and will therefore not always be directly across or along the piece of wood.

Fig. 39 shows a warp typically found in a longcase door. Such warps often develop in the relatively wide pieces of solid wood used in such doors, the problem being aggravated by the differing treatment of the two sides. The back of such a door is virtually bare wood, and

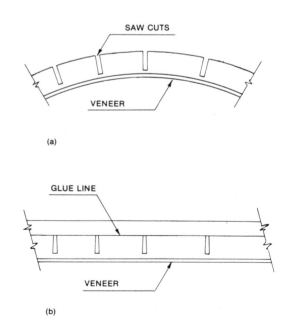

Fig. 37. Correcting a Warp, Using Saw Cuts

therefore absorbs moisture at a different rate from the front, which is either stained and shellacked or plated with polished veneer.

If the door in Fig. 39 is laid flat on its face on a bench, corners B and D would be touching the bench and the other two corners A and C raised off it. To correct the warp, make a series of parallel saw cuts in the back in the direction BD. Three are shown, but five or six may be necessary to avoid any suggestion of their presence in the veneer. The cuts can seldom be made on a table saw because of depthing problems, so they must be sawed by hand; use of chain drilling or a router also deserves consideration. After soaking and flattening, wedge the cuts as in Fig. 38.

If the warp were reversed, with corners A and C below B and D (the door still being face down), make the cuts as before and allow them to close a little when the door is flattened out, as in Fig. 37. As there is nothing to keep the weakened door flat, it is necessary to stabilize it with battens as shown, or, better still, to level the surface and veneer it, the veneer grain running across the door in the direction AB. If battens are used, they should be about ¾" × ⅛", and inset into grooves for neatness. Make them of the same wood as the door, and fill in the narrowed saw cuts with glued, easy-fitting wood strips.

While wood battens are suitable, it is better to glue a thin sheet of wood over the entire back of the door, if total thickness allows. Veneer would not be strong enough, and slightly thicker solid-wood sheet is unobtainable, ⅛" or ¼" flat plywood may be used.

Warps in the swan necks and other places can usually be corrected as described above and fitted with small supporting battens which are normally out of sight. In all such corrections, it is important that the saw cuts share the correcting bend equally between them, so that the veneer goes undisturbed and does not show any localized bend-lines.

A warp in the door of the hood can take a surprising amount of time and trouble to repair. The construction of the arch is far from strong, and an oak arch can bend enough, over a period of time, to crack the glass. The arch may take on a marked backwards lean; this must be corrected by making horizontal saw cuts on both sides and wedging them in the manner of Fig. 38. Note that the wood grain usually runs horizontally across the arch.

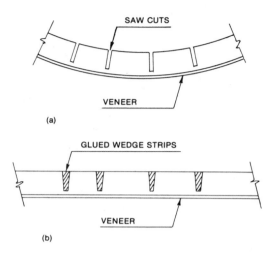

Fig. 38. Correcting a Warp, Using Wedged Saw Cuts

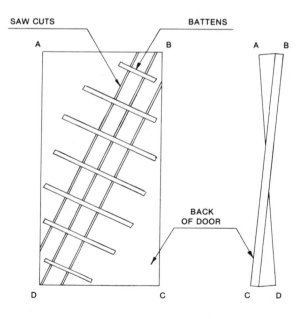

Fig. 39. Saw Cuts in a Warped Door

Twists in the sides of the door are dealt with in the same way, but the cuts must be angled as in Fig. 39; crosscuts would leave the side much too weak.

In a warped hood door, the glass is frequently already broken; if not it may have taken a set over the years, while still resisting the full thrust of the warp. It is therefore far from easy to true a glazed door without inadvertently breaking the glass, so it is best to remove it beforehand if possible. This in itself takes great care, as the old grout or putty is usually almost rock hard, and both glass and veneer are easily broken when the recess is being chiselled out. Work on the hood door undoubtedly requires painstaking and unhurried care.

Structural Soundness

Most of the strength possessed by a longcase comes from glue blocks; the old casemakers used nails sparingly. The old, tapered square nails do not hold well anyway, and often have worked loose. Modern spiral, ring and coated nails hold much better. Regardless, the hammer must be used as little as possible on these fragile cases, for it is easy to knock two pieces off for every one put back on.

It is best to start at the bottom of a longcase and work upwards, so that the case always stands truly upright. If the base happens to be loose, it is tempting to make it rigid by nailing a stout board across the bottom of it. However, this would be a mistake: Longcases were not usually fitted with a bottom originally, because from time to time the cable breaks and a 14-pound weight crashes down. This would smash a bottom board to pieces, so it was left out, thus allowing the weight to hit the floor instead and so cause less damage to the clock.

Small, roughly triangular cornerpieces can be used in conjunction with wooden bars and glue blocks to build up and strengthen the base. New feet may have to be made, and the whole assembly firmly secured with glue and wood screws. As the base is not bound to be accurately made or quite square, it is essential to check from time to time that the case stands vertically while you build the strengthening frame and repair the feet. Not all clocks have four feet; some have only two, at the front so that the clock leans against the wall. With this in mind, it is a good idea to make the rear feet a trifle shorter than the front ones. In any case, the feet usually have to be shimmed when the clock is set up to run. This repair work will be out of sight under the base, but if any should show—say, beside a foot—dark stain will effectively hide it.

The back of a longcase usually consists of two or more full-length pine boards glued together edgewise. As the boards are relatively thin, these joints sometimes come apart. Below the pendulum bob, an internal batten ¾" × 2" can secure the split; it should run horizontally right across the back and be fastened to the sides with glue blocks. Above the bob, it is best to put slightly thinner battens on the outside of the backboards; if put inside, they would probably interfere with the pendulum.

If the backboard has pulled away from the side, tap in a few slim nails, about ¾" long, to hold it in position, and then fit several glue blocks, spaced about 3" apart, inside the case to achieve the required strength. Glue blocks are much more effective than nails alone. If a glue block has fallen off, scrape its seat clear of old glue and replace it with a new one; it is hardly worthwhile to clean up an old block. Brown glue must be used unless the block is being installed in a fresh position.

The seatboard rests on the upper ends of the waist sideboards, where they extend up into the hood; these supports are often somewhat thin, uneven, or out of line with each other. The seatboard may be warped or a little on the short side. All or any of these factors can combine to make the seatboard assume an uncertain or tilted position; if this is not corrected, the cables may slip off the drums or pile up on the clicks during winding.

With the case upright and level in both left-to-right and front-to-back directions, put the movement in place and check that the dial is positioned correctly in the aperture of the hood. It may be necessary to adjust the height of either seatboard support. Note, however, that the dial is not always square with the movement; since the dial takes precedence, the seatboard may have to slope a little, by means of shims, to accommodate it.

When the proper height of the seatboard supports has been determined, make sure that the seatboard sits flat upon them at each end. If the supports are crumbling or in any way doubtful, a short length of so-called 2″ × 4″ wood will strengthen them. These two lengths of wood should run from the front posts to the case back, and should be temporarily clamped in place. A small amount of glue is applied only after their correct positions have been found and marked.

These two individually fitted blocks are then fastened to the existing supports either with wood screws or nails—without, however, the use of a hammer. Drill clearance holes in the supports for several nails, positioning them for maximum strength, and then wind the nails in with an old screw clamp which has had its swivelling foot removed. It is possible to use a normal clamp instead, but it is awkward to wind it inside the case. A small hollow drilled in the end of the clamp's screw can help stop it from slipping off the head of the nail. The clamp can be used in other places where hammering is undesirable.

Longcase Brasswear

Brass case trimmings, such as waist-pillar capitals and pillar-flute rods, are found chiefly in the older clocks that have brass dials. White-dial clock brasses usually consist of only the hood-pillar capitals, the waist-door hinges, and perhaps a hood-door knob. Whatever brasses are present, they should all be removed and cleaned, polished, and lacquered separately. Never apply metal polish to brasses on the case, for it will only result in an unsavory buildup of verdigris.

Veneer

The veneer on longcase clocks can be 0.12″ or more in thickness, which is much thicker than modern veneers. Being an external surface covering, veneer tends to receive any knocks, dents, and scratches to which the clock is subject, and as a result small pieces sometimes break off. All pieces of loose veneer which have fallen off the case should be retained; it is best to clean up both the piece and its seating and replace it without too much delay, using brown glue. If this is done carefully, there is no need of any sanding or other patina-destroying levelling which would make the repair more obvious.

If pieces are missing, then the seatings must be cleaned up, and new pieces of the same wood cut out and fitted. This often entails cutting off ragged edges to form straight mating edges which can more easily match those of the new insert. Always try to pick a piece of veneer with the same grain pattern as the original.

Fig. 40 shows a typical example of missing veneer. The missing section ABD might be at the corner of an American ogee or the base corner of a longcase. The jagged line BD is too difficult to match, so cut the veneer along the dotted line BC and remove the area BCD. Now remove all remnants of old, dried glue from the area ABCD, and cut a piece of identical wood to the same size and shape; thickness and grain direction must match those of the original wood. In Fig. 40 note that the grain runs parallel to BC—a typical pattern—which unfortunately makes the corner at A somewhat fragile.

Cut and try out the new insert until it fits snugly. Then set it in place with brown glue, removing any excess. Lay a small piece of cardboard (or something similar) on top, fol-

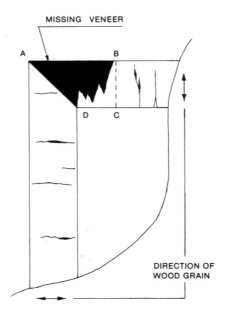

Fig. 40. Repairing Flat Veneer

lowed by a small block of smooth pine, and clamp all this together; allow several hours for the glue to set. The cardboard stops the block from sticking to the veneer, which could cause damage on removal. Any sticky remains of the cardboard and surplus glue can easily be removed with a damp rag.

Fig. 41 shows how to repair veneer that lies on a curved surface, a configuration often featured on school and shelf clocks. The area PRSU in Fig. 41a represents missing veneer;

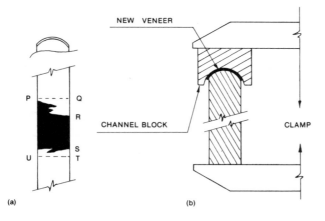

Fig. 41. Repairing Curved Veneer

the ragged edges must be trimmed to the lines PQ and UT to enable a new insert of veneer to be fitted. Because the grain of the veneer is parallel to PQ, the edges tend to lift somewhat, and are then easily damaged.

The bend radius is relatively small, so it is essential to cut the insert to size and pre-bend it. This is done by wetting the insert with very hot water, and carefully bending it around a piece of dowel, or metal rod, binding it in place until it dries out. The radius of the dowel must be a little less than that of the job, because the insert will relax slightly when taken off the dowel. It is best to repair small sections at a time, binding each of them to the dowel with masking tape. The joints at PQ and UT are not hard to hide, since they lie with the grain; however, if the distance PU is longer than an inch or so, there is a good chance of cracks developing during the bending process.

In some instances, it may be possible to hold the insert to the job with masking tape until the glue dries, but better results are obtained by using clamps and a small block with a smooth, curved channel in it, as shown in Fig. 41b. The channel can be gouged out of a small pine block and then smoothed with sandpaper wrapped around a piece of dowel of appropriate diameter. The channel must be a good fit onto the job, or the veneer will bond unevenly, allowing gaps to form.

When the glue has set and the clamps have been taken off, sand the new insert lightly until it is level with the original. Chisel off any excess and sand the edges smooth. This should be done without delay, because projecting veneer invites accidental damage. Finishing, the next step, will be dealt with later in this chapter.

School Clocks

School clocks and other wall clocks often exhibit loose segments in their octagonal tops and wooden bezels; the glue has powdered away, and individual pieces have separated. The remedy is to take out each loose piece, file

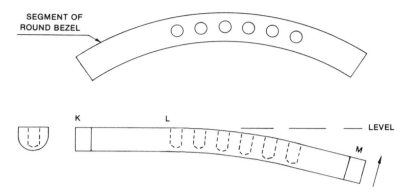

Fig. 42. Correcting a Warp, Using Holes

and scrape off all the old glue, and put everything together again using brown glue and clamps. A strong, surrounding cord, pulled tight, can also prove effective. Make absolutely sure that the octagon or bezel is set up flat, for with so many glued joints a warp can be very troublesome to correct.

Fig. 42 shows a warped segment LM of a round wooden bezel KM. It is not desirable to use saw cuts here (as shown in Fig. 37), because they would show on both sides of the segment. However, drilling a series of fairly large holes will solve the problem very tidily. The holes must be quite close together so that the wood is uniformly weakened. Pour hot water into them, allowing it to soak into the wood for about twenty minutes. When the segment is relatively amenable to straightening, it can be clamped between two birch boards, preferably in a somewhat overstressed position. This can be achieved through the insertion of small, well-placed pieces of wood, but twisting must be avoided.

When dry, the wood is glued and clamped back in its correct position. Any residual moisture will not give trouble (brown glue can be thinned with water); it will simply take longer to set. As with saw cuts, the use of holes and hot water enables straightening to be achieved without refinishing.

Vienna and German Wall Clocks

These cases typically consist of a fairly solid top and base joined together by a backboard and two slim, picture-frame sides. The glazed door constitutes a third picture frame, and the whole assembly is far from strong. In fact, there is so little wood in the side frames that, should they loosen, there is no real purchase for a screw or nail. Aside from the backboard, only the front members of the side frames hold the top and base together, the glass being too loose to provide any rigidity. When their already weak joints loosen, these cases often have to be completely taken apart to repair them. Glue blocks may be added to strengthen the joints, but brass brackets, screwed to the sides at top and bottom, offer the best support. Being inside the case, they can be seen only when the door is opened, and are unobtrusive even then. They should be of 0.06″-thick brass, about ¾″ wide × 3½″ long. Two screws hold 3″ of the bracket to the side, and the remaining ½″, bent at a 90° angle, is screwed to the base or top.

Although the use of metal brackets may be frowned upon in certain circles, it is essential that the front members be strengthened, particularly since one of them has the additional duty of supporting a rather heavy door. In this type of clock, case size varies considerably, but the same small, weak hinges seem to have been used on all of them. These hinges should be checked periodically, as they sometimes bind. Binding imparts a continual back-and-forth motion to the front member, and so slowly loosens it. It is a good idea to replace such hinges with larger ones.

Due to a lack of case rigidity, the glazed doors of these clocks usually end up binding against the top or base, and so are difficult to open and close. Owing to variations in humidity and temperature and the general looseness of the case, a fairly generous clearance must be provided. Sanding is inadequate to deal with this, and planing can easily take slivers out of the capitals or blocks of the door pillars, so it is best to feed the top and bottom of the door very slowly across a hollow-ground, veneer-cutting blade on a table saw. A cut of about 0.05" generally proves sufficient.

Warps and twists in these flimsy cases are not easy to remove. Angled saw cuts in the door frame filled with glued strips can be used to take out twists (see Fig. 39), and a series of holes can be used to straighten a curve (see Fig. 42). Hot water used alone is not adequate; the door will have to be refinished, and the warp may return later, for there is nothing to keep such frames straight and true.

A curved door frame may have created a set in its glass, and by reversing the glass, this set can be used, within limits, to oppose any tendency in the straightened frame to return to its curve. It is good policy to slightly overcorrect all warps, anyway, and the reversed glass will generally fit in quite well. The idea is that the wood will still possess enough strength—but only enough strength—to overcome the over correction, and will finally stabilize when straight. This small movement will be opposed by the glass, but it usually occurs slowly enough to allow the glass to accommodate.

When the case is ready for refinishing, try out the movement in it, and check that the glass of the closed door does not touch the minute hand or its arbor, as even light pressure can stop the clock. This point is often overlooked.

Surface Treatment of Wooden Cases

The surface finish on most old clocks is either shellac or some kind of varnish. If the case is in poor condition, or has been painted or heavily waxed, it is best to strip it completely, repair the case where necessary, and refinish it with stain and a good grade of varnish.

Stripping

There are many different brands of paint and varnish stripper, but all of them deserve careful handling. (Rubber gloves will protect delicate skins.) Anything capable of dissolving old, dry paint will certainly not serve as an eye lubricant or skin lotion; toxic fumes make ventilation essential, too.

Most paint strippers come in either liquid or jelly form. The liquid can be brushed onto items which can be stood in a shallow bath, and the jelly is useful in places where the liquid would run away too quickly. Some strippers require neutralizing, many do not, but all residues must be removed or exhausted prior to staining.

It is best to avoid total immersion of the piece, since this may bleach the wood unduly or dissolve glue and cause things to fall apart. Localized applications will usually prevent these hazards. Given this restriction, the stripper must be used liberally, and given time to soften the paint until it can be scraped off or removed with a handful of sawdust. Additional minor applications, and use of the pointed slivers of wood, may be required to clean out the deeper recesses and angles of carvings and pressed wood.

After stripping, the repairs described previously in this chapter can take place. Then lightly sand the case with fine sandpaper. This must always be done along the grain of each piece of solid wood or veneer; cross-grain sanding may appear to be the easy way, but it will show up badly at the varnishing stage. This applies equally to stripped cases and veneer repair inserts.

Remove all wood dust from the case by means of a vacuum cleaner or brush; do not wipe the case with a cloth, because small pieces of lint will get caught in the grain of the wood.

Staining

If the case has been repaired with new wood or veneer, these new areas must now be stained to match the surrounding grain as closely as possible. If the case is being completely refinished it may or may not require staining. Rosewood, for example, can hardly be improved with stain. In most cases, however, staining is advisable, because it tends to even out localized areas of differing hues and enhance grain patterns.

Veneer repairs often require small quantities of two or three different stains mixed together for matching purposes. In contrast, stain for a refinishing job may be taken straight out of the can. This is a matter of choice. If, for example, it is desired to darken the usual yellow oak stain, a dash of walnut can be added to it; always be sure to mix enough to cover the whole clock case, or matching problems could arise.

For most purposes, pint quantities of mahogany, oak, and walnut oil stains made by the same maker are all that is required, for these can be mixed or thinned to give a variety of shades. (Very few old clocks were made of the lighter colored woods such as cherry.) Brush the selected stain onto the case fairly liberally so that it penetrates all corners. The longer the stain lies, the darker the result, so to avoid mismatched areas do not stop halfway down a panel. Give the stain a few minutes to soak in, and then wipe off the surplus with a clean rag. As stain raises the grain slightly, give

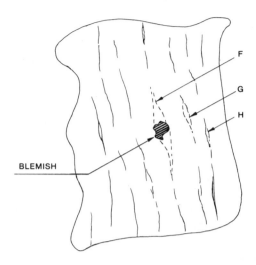

Fig. 43. Faking Up Blemishes in Wood

the case a very light rub with fine sandpaper when it is thoroughly dry.

Faking Up

As the rate at which wood absorbs stain varies somewhat over small areas, a few spots, either too light or too dark, usually show up after stain has been applied. A smear of glue left on the case, for example, would cause a pale spot where the stain failed to reach the wood. This is easily sanded off and restained, but other flaws in the wood often go deep. These must be painted to merge with the surrounding grain. This procedure, sometimes known as "faking up," is a necessary step in all wood finishing; it amounts to artistic camouflage. Only a tiny quantity of paint is mixed, and it is then applied with a small artist's brush or perhaps lightly smeared with a wet fingertip, depending upon surrounding colors and grain patterns.

In Fig. 43, the blemish, be it dark or light, can be disguised to resemble a natural grain formation by painting the area F to include it. Applying the same coloration to two smaller areas, G and H, may be desirable to make the effect more natural. This type of faking is often preferable to simply trying to fade out the blemish area only.

If the wood in Fig. 43 is mahogany, for example, a mixture of red and black would be used to give F, G, and H a slightly darker hue. Black with brown serves for walnut, and yellow plus a touch of black for oak. Over small areas, one or two well-placed and cleverly shaped streaks of blackish paint will make mahogany pass for the surrounding rosewood.

Any small holes filled in by means of wood filler, or a shellac stick, will cause blemishes because of their differing stain-absorbing characteristics. Several paint applications, followed by light touches of wet-or-dry emery paper and water, may be needed before these spots are smooth enough to be faked up satisfactorily. Any differences in gloss between painted and unpainted areas are unimportant, because they will disappear after varnishing.

Final Finishes

In many instances of veneer repair, as distinct from a complete refinishing job, the finish on much of the case is already in reasonably good condition. It is important to find out whether this finish, which may well be the original one, is of shellac or varnish.

To test this, rub a small, inconspicuous area of the finish with a rag soaked in a little methyl hydrate. If the finish softens, it will be shellac; varnish will be unaffected. If it is varnish, newly repaired areas need simply be varnished and rubbed down to match the rest of the case.

Shellac

If the finish is shellac, the whole case can be readily cleaned and brightened by simply working it over many times with a linen rag soaked in methyl hydrate with a few drops of boiled linseed oil added. The oil will promote smoothness, and prevent the formation of the whitish bloom that sometimes appears under certain conditions of humidity and temperature.

Firm and persistent rubbing in this fashion will slowly remove entrenched dirt, leaving a pleasantly smooth finish akin to the original sheen. However, do not allow the rag to dwell even momentarily in one place, or it will stick to the finish and mar it; keep it moving under firm pressure. Shellac dries very rapidly—in a matter of minutes—but it is wise to avoid touching it for several hours afterwards, all the same. If small repairs have been made, there is usually no need to supply more shellac, because as rubbing proceeds the shellac already on the case will spread out over the repaired areas. It may take an hour or two to cover a clock case completely in this way, but the result is similar to that of a complete refinishing job, without the need for stripping.

If for any reason additional shellac is required, it is important to know a thing or two about this commodity as there are certain problems involved. It seems that the lac insect secretes for itself a certain protective coating, and this material is the basis for lac flakes, which are available in brown, orange, or bleached form. The bleached variety is the most suitable for our purpose; the others impart too much coloring.

The term *shellac* is used for both lac flakes and the liquid solution. Liquid shellac is made by dissolving shellac flakes in methyl hydrate or some other member of the alcohol family. The flakes take several days to dissolve, and various strengths, known as *cuts,* are possible, according to the time allowed for assimilation. Only enough liquid shellac should be made to satisfy present requirements; a heaped tablespoonful in a pint of methyl hydrate is usually enough. Shake the mixture in a clear jar as often as possible for at least two days. If all the flakes disappear, add some more and continue shaking until no more will dissolve.

The saturated solution thus obtained must be used within two weeks, if it is to harden properly as a surface finish. This fact is little known or neglected; usually by the time trouble develops, the age of the shellac, if it was ever known, has been entirely forgotten. If shellac much older than two weeks is used as a surface finish, it will not stay hard every day of the year. In the warmth and humidity of summer, it will become tacky—and will continue to do so for decades.

Store-bought shellac—invariably several months old—is generally used as a sealer for internal woodwork, house doors, and so on, and only a thin coating is put on these surfaces. As this is covered by one or more coats of varnish or paint, the softening problem does not normally arise.

Furniture, on the other hand, is still occasionally "French polished," as were many old clock cases. French polish is simply several coats of shellac applied in a particular manner to ensure a smooth, even finish. A small square of linen is folded carefully to form a pointed pad, known as a *rubber*. The shellac is not applied to the rubber directly, but to a small separate rag inside it, from which shellac slowly seeps through to the working surface in a more-or-less controlled manner. Five or more coats are applied over as many days, slowly thickening the coating. French polishing takes far more time than brushwork.

Not long ago, I was consulted about a set of fairly new shellacked dining chairs which in their current condition were unusable. At a dinner best forgotten, the chairs had bonded themselves to the clothing of the diners and done their best to pull the shirts off their backs. The shellac obviously was somewhat elderly when it was applied; the chairs had to be stripped, stained, and varnished.

Shellac is always vulnerable to alcohol. Rings caused by imbiber's tumblers are not easily removed; the whole surface may have to be redone. Shellac therefore serves better on clock cases than on other types of furniture, but it can never be as serviceable as varnish.

Varnish

A good varnish is probably the best all-around final finish for woodwork. Oil varnishes are available in gloss and semigloss form, and may be applied by brushing, rubbing, or spraying. My own preference is to use a fine brush and a good-quality, oil-based satin varnish. Three evenly applied coats, each rubbed down lightly when dry, give a clock a pleasant glow which can be polished to come quite close to the original patina. Occasionally, a clock with a lot of inlay may look better with a gloss varnish. A medium lustre can also be obtained by mixing gloss and satin varnishes together.

Wax

Many people like to polish their clock cases from time to time, and there are many preparations available for the purpose. Unfortunately, prolonged use of some of them results in a marked buildup of wax, and eventually this may crack off in places, leaving a pock-marked surface.

Wax is undesirable as a surface finish for other reasons—such as its being soft and dirt-collecting. It offers little or no protection, and is difficult to remove when a more durable finish is desired. In view of these characteristics, I cannot recommend its use as a finish or polish.

Cleaning and Polishing

An effective cleaning and polishing mixture, which is used by antique-furniture dealers and, apparently, some museums, is easily made from commonly available liquids. Its smell is inoffensive—even quite pleasant. It is simple to apply with a soft cloth, and leaves a soft sheen without any undesirable buildup. The mixture must not be allowed to remain on the surface—it must be polished off straight away. This will avoid any chance of its softening the finish (shellac, for example). Because the components tend to separate when the container is left standing, vigorous shaking is required before use. The formula is as follows:

Boiled linseed oil	½ pint
Cider vinegar	½ pint
Turpentine	½ pint
Methyl hydrate	1 tablespoonful

Glass Domes

Skeleton clocks and other eye-catching timepieces are usually housed under glass domes, which must be raised frequently to allow winding and adjustment. Oval glass domes in particular are difficult to find and need careful handling. It is all too easy to break one when lowering it over a clock; the edge generally hits the brass of the movement, chips or cracks appear, and the dome is ruined. Other forms of accidental damage occur, although less often. Because the edge of the glass is the area most likely to be damaged, it is wise to safeguard it.

Strip grommet is used to protect electric cables or hoses where they pass through a hole in a metal bulkhead or panel. It is a flexible strip of nylon or similar plastic that resembles two closely spaced rows of miniature castellated battlements, which allow it to fit around the edge of a hole. It can be ordered from electronics suppliers in several sizes, the most usual colors being natural nylon and black. It is sometimes known as caterpillar grommet.

To protect the edge of a dome, select a length of strip grommet of the smallest size that will embrace the edge of the glass and fit it on (see Fig. 44). If necessary, use two pieces to encircle the edge of the dome, cutting them so they butt together. When they fit snugly without gaps, remove them, run an even line of epoxy resin along their centerlines, and then bond them to the edge of the dome, using strips of masking tape to secure them in place.

Some domes fit into a groove in a wooden base, and it may be necessary to widen this groove slightly to accommodate the width of the grommet. Although this increase in width should be small, it is essential to maintain good clearance, for the wood may warp or shrink—and the glass dome will do neither.

Thus placed, the strip grommet absorbs the inevitable small shocks that might otherwise chip or crack the dome. It is well worth installing.

Marble or Slate Cases

These heavy cases are not as strong as they look. They are usually held together with plaster, and even the vibrations caused by pulling a case across a bench can cause a panel to come off. Replacements for these pieces will be very difficult to obtain, so it is best to keep the cases all in one piece. When carrying them, always hold them from underneath; in fact, it is wise to carry all clocks that way.

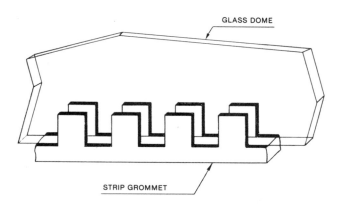

Fig. 44. Glass Dome Protection

IIII

GLOSSARY OF HOROLOGICAL PROCEDURES AND DEVICES

Anvil: A small, hardened-steel block measuring 1¾" × 1¾" × ¾". Several shapes are available, the hexagonal and rectangular being the most common. All have several holes of various sizes, and often a few slots of differing widths; such anvils are very useful in bushing and many other clock procedures.

Arbors: clock shafts that carry gears, lifters, etc.

Barrels: *Hook Repair.* When a mainspring breaks, the released force is usually enough to drive the mainspring hook outward, thus loosening it somewhat. From the outside, the shape of the protruding hook then resembles a tiny, steel-topped, brass cone, whereas normally it is flush with the outside of the barrel. The hook must not be simply hammered back in again; this will either loosen it further or knock it out altogether. The essential thing is to force the brass cone back into place, and it will then tighten up again around the steel hook. This is done with a hollow punch, using a suitable vise-held support (See Fig. 45). An ideal support would be a 5"-long, ½"-square

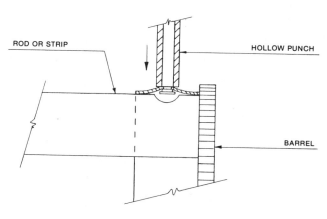

Fig. 45. Tightening a Loose Barrel Hook

steel bar or strip—or a ¾"-diameter round rod. Clearance for the hook must be provided in its top side and any sharp edges must be rounded to avoid damage to the barrel.

Bench key: See **Let-down key.**

Bending tools: Fig. 46 shows two useful bending tools which are easily made from ¼" mild-steel rod. They permit lifter adjustments in places that are inaccessible to long-nosed pliers. The rods should be about 7" long, and should be slightly tapered prior to bending.

Bim bam: This is a form of striking applied to both hours and quarters: Two tones in sequence, usually a high, then a low; one pair for each quarter hour or each hour struck. When the quarters are struck as one, two, or three pairs of notes, the hour is struck on a single gong to avoid confusion.

Bluing: Heat bluing is applied to steel parts such as screws, hands, arbors, etc. It is purely a surface finish, and looks best on smooth, finely finished parts. The parts to be blued must be thoroughly burnished to a bright finish; otherwise the color will not be seen during the heating process.

The parts are not heated directly, as this provides insufficient control, resulting in an uneven finish. The recommended medium is a bed of brass filings, but, lacking this, a shallow tray of sand can be used instead. In either case, it is essential to apply the heat evenly over the full area of the tray, so that a long piece, such as a longcase clock hand, develops the desired depth of blue simultaneously along its entire length. A part to be blued is laid on the sand and then pushed gently down until its top is flush with the sand surface. A 0.050"-thick brass tray about 7" × 2" × ½" serves for most jobs, the heat being applied to the underside. Do not, by the way, use an old aluminum egg-poaching pan; the bottom will burn out and the sand will drain down into the cooking range. (I prefer not to elaborate on just how I came to find this out.) Use a propane torch in the workshop, and be ready to pick the item out quickly when blued.

While sand works well, it does take time to warm up, and I have found that a strip of 0.062" brass about 1" wide and 9" long works just as well and is somewhat faster. Holes must be drilled in it to accept screws, and clock hands must be flat in order to contact the strip in as many places as possible. The flame is played back and forth beneath the strip until the desired color is obtained; the strip is then turned over, letting the part fall off, sometimes into a dish of oil. This is much easier than trying to pick out several pieces at once from a sand tray. (If they are not removed instantly, they will overheat and the blue will fade to grey.) Should this happen, simply brighten the parts again and repeat the process. The type of oil used is not critical.

Bluing slows rusting only marginally, so it is a good idea to lacquer steel clock hands, back and front, after bluing them. In addition to

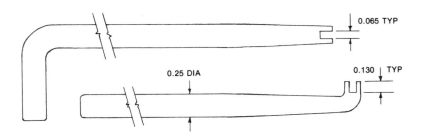

Fig. 46. Bending Tools

providing protection against rusting, lacquering enhances the appearance of blued hands, longcase hands in particular.

It must be noted that the requisite rich, dark purply-blue color occurs when the steel reaches a temperature of about 550° F. As this is appreciably higher than the 400° F at which most soft soldering is done, it is not feasible to heat-blue parts that have been soft soldered. As silver soldering requires some 1500° F of heat, its strength is unaffected by heat bluing.

Color dyes are available for parts that cannot be heat blued.

Bushing: *Barrels.* The bearing in the end of a barrel extends into the barrel to provide a greater length of bearing surface than the thickness of the end by itself would allow, and the height of this extension must be measured. This is easily done with the depth gauge of a vernier caliper. When this measurement is known, the next step is to enlarge the worn hole in the lathe until it will accept a bushing. As this hole must be concentric with the pitch circle of the gear, it is best to grip the barrel by the gear teeth rather than by the barrel itself. Naturally this must be done with care, or damage will result.

A good way of mounting the barrel in the lathe by its teeth is to mount a piece of hardwood, roughly 6" square and ¾" thick, on the lathe faceplate, and then turn out a spotface in the wood deep enough to take the gear with a firm press fit. Since many barrels have four large holes in their ends, the barrel may even be screwed to the hardwood through them, as insurance. The worn hole is now turned out until its diameter exceeds that of the accurate, polished arbor's bearing section (pivot) by about 0.100". Do not use a drill, because if it should grab it may pull the barrel out and damage it. For the same reason, make one-way cuts from right to left of not more than 0.003" thickness.

Another way is to scribe the rough outline of the hole-to-be, using an adjustable square. Rest the teeth against the body of the square and set the rule so that its end is tangential to the required hole. Scribe this short tangent in five or six positions, some 72° or 60° apart, rotating the gear accordingly, to mark a polygon in which the hole circle could be inscribed. Now mount the barrel, across its flats, in the vise, using appropriate protection against scoring, and file out the hole almost to the tangent points. Keep the hole as circular as possible by using successively larger round files.

Using the outside jaws of an accurate three-jaw chuck, mount the barrel in the lathe, with the jaws gripping the teeth, not the barrel. Make sure that the barrel is fully inserted against the chuck jaws, check that no light shows at the three contact points, and then tighten the chuck—moderately. *Do not screw the chuck up tight as though you were gripping solid rod!* As each jaw covers several teeth, the gear will be positioned quite accurately.

Now turn out the center hole to the required size, taking light cuts of 0.002", and cutting only on the right-to-left traverse in order to obviate any chance of a pullout. The cuts are minimal, but the job will not take many minutes, as there is only a small amount of metal to be removed. Once again, do not be tempted to use a drill: With this setup, a grab by a sizable drill would very likely damage the teeth of the barrel beyond repair. Lightly chamfer both sides of the hole to remove burrs and facilitate setting and riveting the bushing.

It is best to turn up bushings for barrels and barrel covers as they are required, rather than attempt to buy a stock of "possibles." The diameter of the bearing sections of typical German barrel arbors is about ¼", so a bushing will need to be about ⅜" in diameter with a flange of up to ½" O.D. A total length of some 3⁄16" will be needed. The flange is placed inside the barrel, and its thickness determines the degree of end play; however, this can be adjusted slightly by dishing the barrel end or its cover, as shown in Fig. 77 (see **Mainsprings in barrels**).

Now turn up a bushing from a piece of ½"-O.D. brass rod. Center the hole carefully, and drill it out to be about 0.007" less in diameter than the arbor pivot (see under **Drilling**). Fig. 47 shows a bushing, with typical dimensions. Dimension F is made to suit the end play, but is basically equal to the height of the original bearing extension as measured earlier, plus a few thousandths of an inch to compensate for wear.

To preserve concentricity, the bushing must be a press fit of at least sufficient tightness to keep it from being shaken out. Face off the ends of a few inches of ⅜"-diameter, hardwood (e.g., birch) dowel, and use it as a punch to tap the bushing into place. As most anvils do not have a 0.50" hole to clear the emerging 0.375"-diameter bushing, the inner ring of a discarded ball bearing will serve very well instead. It is a good idea to keep a selection of such inner and outer rings, together with a few balls; all these very useful parts are accurately made of very hard steel. The balls come in handy for dapping operations, such as curving a hand washer, but they must not be struck directly or they will damage the hammer face.

Note that if a burr had been left around the hole, a gap may well occur around the bushing, at point G in Fig. 47. See to it that no such gap exists, and then rivet the bushing firmly in place all around, as indicated at H. Mount in the vise a stub of 1"-O.D. steel rod, with its ends faced off flat and smooth, where it will serve as an anvil inside the barrel during riveting. The end of the stub must be well finished, as otherwise it will mar the flange of the bush. (Being an end bearing, its integrity must be preserved.)

Now broach out the hole to suit the arbor, using the concept of Fig. 48 as a general guide. (Note, however, that Fig. 48 is for a smaller bearing in a plate, and the dimensions differ.) As the present bearing is more than twice as long as that of Fig. 48, and the pivot-to-pivot distance of the barrel arbor is again about 1½", the circle described by the free end will not much exceed 0.10" in diameter to give a satisfactory clearance. The geometry of Fig. 48 is useful for working out clearance for other bearings, as required. The important thing to remember is that the dimensions of both bearing and arbor must be taken into account in each case, when determining the diameter of the circle described by the free end.

Break the sharp edges of both sides of the hole by giving them a light twirl with a hand-held countersink.

A barrel cover is bushed in the same way as a barrel. A cover will mount easily in the outside jaws of the lathe chuck.

Assemble the arbor, barrel, and cover, and adjust for end play as shown in Fig. 77 ("Dishing a Barrel"). The flanges may be remachined, if necessary, since here concentricity is not involved.

Bushing: *Plates.* Pivot holes in plates for no. 1 arbors are bushed in the same way as barrels and barrel covers, the flanges being generally outside the plates.

Smaller holes, which are in the majority, require no flanges; they are best lightly riveted both sides. Assuming clean, accurate pivots, the amount of wear present in the pivot hole must be assessed in order to determine whether bushing is required. Experienced eyes can see instantly what is needed; until experience takes over, it is important to have a quick, practical guideline for use.

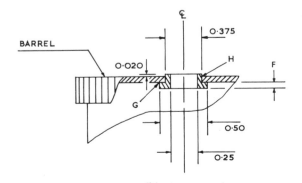

Fig. 47. Bushing a Barrel

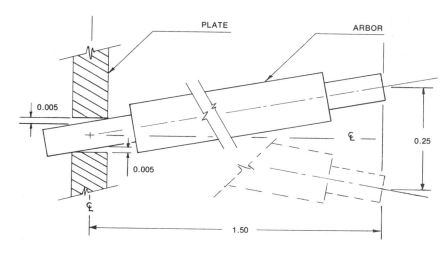

Fig. 48. Bearing Play

A bearing that is correct for a clock would be too loose in most engineering applications. A clock bearing must not be too precise; otherwise it will "seize up" when the oil thickens, usually after some months of running. As a rough guide, the pivot hole in a clock plate should be about 0.005″ greater in diameter than the pivot which turns in it. There is no need, however, for expensive equipment to measure this clearance, as sufficient accuracy can be obtained from the method described below (see Fig. 48).

If we assume a plate thickness of 0.060″ and a plate-to-plate gap of 1½″, play of 0.005″ will allow the free end of the arbor to describe a circle of ¼″ diameter, as shown. As a guide, if the circle diameter much exceeds ¼″, the pivot hole needs bushing. The circle diameter can be measured quite accurately by clamping the plate in a vise, inserting the pivot, and swinging the free pivot to its limits over a steel rule. This must be done in at least two positions, 90° apart, for a worn hole will be roughly oval, and the free pivot will not be able to describe a circle. For example, play may well be 5⁄16″ across the North-South diameter, and ½″ or more across the East-West diameter.

Having determined that bushing is necessary, the exact method of doing it must be decided. There are several bushing machines on the market, some of them costing several hundred dollars. It would seem that plates can be bushed without even taking the clock apart. However, I cannot endorse this idea, for several reasons: I do not see how the pivots can be trued and polished, and other repairs effected, without dismantling the clock. A clock that is worn enough to require bushing invariably has other faults, and these can hardly be corrected without pulling the plates apart. There is also the question of possible tooth repair, a thorough cleaning, rust removal, polishing, etc.

In addition to the high initial cost of a bushing machine, there is the expense of maintaining a stock of suitable bushings, many of which will seldom be required. Therefore, unless some form of mass processing is involved, I cannot see that ownership of a bushing machine is at all necessary. As will be seen, bushings can be made on the lathe and installed accurately and conveniently using a drill press and hand tools, so that each bushing is a custom fit for its pivot.

Although bushings can be purchased, they are offered in assortments, and a large percentage of them may be unsuitably small or short for the work at hand. It seems better, therefore, to make bushings of the types most used, from the brass bushing "wires" which are available from suppliers. These are really small, thick-walled brass tubes about 3″ long. Unfortunately, assortments are again the general rule, but tubes that are not used to make bushings may well serve other purposes—for example, as a pipe for a second hand. (See below under **Bushings:** *Making*, for details of how tubes are cut into bushings.)

In the case of a badly worn oval hole, scribe two crosslines to theoretically pass through the true center of the pivot hole, and then file out the hole until it is centered on the crosslines, as shown previously in Fig. 20a. Most holes are not worn to this extent, but, even so, the largest available cut bushing may not be quite large enough, or if it fits its hole may be too large for the pivot. In the latter case, rivet a solid plug of brass in the hole, join the crosslines again, and drill a new pivot hole where the lines cross. The plug is fitted in the same way as a normal bushing.

The wall of a bushing must be substantial, but its thickness is not too critical. As a rough guide, the wall thickness should be equal or nearly equal to half the diameter of the pivot; this means that the O.D. of the bushing is almost twice the diameter of the pivot.

Let us take as an example a pivot which, when trued and polished, has a finished diameter of 0.058″, a typical figure for an American clock. The requisite bushing would then need a wall thickness of 0.029″ or thereabouts, and its O.D. would be 0.116″. The in-hand brass-tube diameters closest to this figure happen to be 0.117″ and 0.108″. Either figure would serve; even the next size down, 0.104″, would be adequate. In this case 0.108″ is perhaps the best choice; allowing for clearance, a wall thickness of about 0.022″ will result, assuming a clearance of 0.005″, as shown in Fig. 48.

In most cases the pivot holes are not worn to an extent that would cause the true center of rotation to wander appreciably. Such holes, although enlarged, can simply be drilled out without prior filing. For a tight fit a 0.108″-O.D. bushing requires a hole of 0.107″. A no. 37 drill (0.104″ diameter) would be quite suitable, allowing for the fact that it will probably cut a hole about 0.002″–0.003″ oversize. Another approach is to use a no. 38 drill (0.1015″ diameter), followed by a no. 36 (0.1065″). After drilling, chamfer both ends of the hole lightly with, say, a no. 10 drill, to remove any burrs and provide riveting space. Do not neglect this step; all too many people try to rivet something without providing space into which the metal can move.

Now set the clock plate, either side up, on a solid, hole-free part of the anvil, and set the bushing over the pivot hole, with its parting-off spur underneath. Due to the natural perversity of inanimate objects, the bushing will most likely position itself as shown in Fig. 49a. A push from side A will not prove useless, but a gentle nudge from the end of a flat punch on side B, as shown in Fig. 49b, will upright it immediately.

Now tap it down using the flat punch, as shown in Fig. 49c. Do not attempt to drive the bushing home with the hammer itself; as a direct hammer blow is usually inaccurate, it can quite easily ruin both bushing and the hole. When the bushing is down fully, then and only then use the hammer to form a small mushroom on the bushing, as shown in Fig. 49d.

Now slide the plate on the anvil so the bushing is over one of the anvil's holes (one that easily clears the bushing), and hammer the bushing almost flush with the plate surface (Fig. 49e). Now turn the plate over and do the same to the other side, riveting it as depicted in Fig. 49f. File both sides flush (Fig. 49g), using a fine file (00 grade) so as not to score the plate.

The pivot in our example has a diameter of 0.058″, so drill out the bushing hole with a no. 55 drill, which will give a diameter of about 0.054″. Now, using a suitable five-sided tapered broach, enlarge the hole to the configuration shown in Fig. 48. As you broach the hole, keep checking that the broach is normal

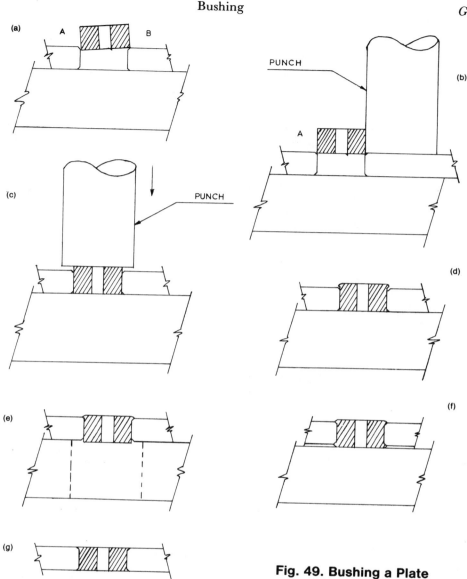

Fig. 49. Bushing a Plate

to the plane of the plate. Do this by eye, looking at the plate edgeways from the side, then turning it 90° and looking at it edgeways from the top. Alternate these positions and make sure that the broach is always perpendicular to the plate in both positions. Broach from either side of the plate, and keep rotating the broach as you withdraw it; do not stop suddenly and withdraw it, or small ridges may be left inside the hole. Smoothing broaches (with oil) may be used at this stage, but they are not essential.

Try the pivot in the hole until entry is obtained. Do not use force, or the steel will pick up brass and cause a jam. Oil may be used, but it will hold swarf, which, along with the oil, will have to be removed at each entry trial. With the pivot in its hole, set a finger on the free pivot at the other end of the arbor, and circle it around. Describe an imaginary circle on the other plate with the free pivot; this circle must have the pivot hole in that plate as its center. This means, of course, that the arbor must swing to either side of the perpendicular by equal amounts, and must do so no matter what edge the plate is regarded from.

Basically, if the remote pivot hole lies not at the center but close to or actually on the circle thus described, then the arbor will bind and refuse to turn. The same conditions apply to all such arbors.

When hole clearance is satisfactory, take a larger drill—say, a no. 35—and twirl it lightly between the fingers so as to break the sharp edges at both ends of the hole. This point is frequently given little consideration; yet, unless it is done, I would not reassemble the movement involved. Oil does not move easily over sharp edges; the small chamfer applied to the outside of the hole tends to act as a very small oil sink. The inside-edge chamfer, however, is the most essential of the two, for a sharp edge, even one without a burr, can seize on any small radius at the base of a pivot. Even though every care be taken to reduce such a radius, it will always be present, however tiny, simply because the tip of a lathe tool cannot be made infinitely sharp. It must be stressed, however, that these two chamfers are small; they are not in any sense deep enough to be called countersinks.

Sometimes the arbors of a clock are slightly short, and therefore have too much end play (see under **Pivots** for how this may happen).

Since in this situation only one or two arbors are usually at fault, it is not feasible to close the plates up a little; this would jam the full-length arbors. Localized warping is poor practice.

There are several remedies, a new arbor being the most obvious; however, it is generally acceptable to fit a bushing that projects sufficiently from the clock plate to take up the slack. This can even be done at both ends, if meshing requirements so dictate. The pivots may be long enough to extend past the outside of the clock plate, but if they do not, the configuration shown in Fig. 50 must be used.

Note that a flanged bushing is used here, because it can be riveted. Some clock repairers take a short cut and simply tap a normal, straight bushing part way into the plate. This is poor practice, however, as the bushing is not really secure. It is all too easy to insert the bushing off true, because it has such a short grip, and it may come out during drilling or work loose later on.

Turn up the flanged bushing on the lathe, and drill the relief hole before parting it off. It is best to lock the tailstock when drilling such holes, so that the drill cannot grab; this is also the reason for not drilling it out after riveting it to the plate. Lightly chamfer both ends of the hole in the plate, as usual, and make the bushing a press fit as described previously. Rest the flange on a smooth, hard surface and swage the end of the relief hole outward with an oversize center punch. A punch that has become too thick for normal use is very useful for such jobs as this. Broach the hole out in the same way as a normal bushing.

American-style movements often have external escape gears, and since the bearing in the escape cock is one of the most critical in the whole clock, it is essential to bush it if even the slightest excess play is evident. Most escape cocks are brackets of a keyhole shape. If a worn pivot hole is drilled out in the usual way, in a drill press, there is a danger of twisting the bearing disc off the neck of the keyhole if the drill grabs. To obviate this, grasp the bearing

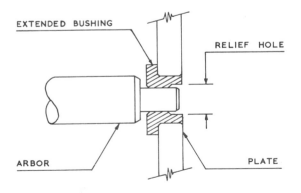

Fig. 50. An Extended Bushing

disc itself in the drill-press vise, making sure that if grabbing does occur the drill can neither pull the disc out of the vise nor jerk the vise itself upwards. A pair of heavy pliers may also be used, but the above method is best.

Press the bushing into place in the normal manner. Use a small flat punch to rivet the inner side, and a large flat punch as an anvil when riveting the outer side. Make sure that the cock is parallel to the plate, and that arbor end play will be correct, before drilling and broaching the bushing. Lay the arbor across the clock plates to assess end play.

If the bearing disc at the end of the escape cock is unduly small, it may crack open as the bushing is pressed in. In this case it is best to soft solder the bushing neatly; all traces of solder, save that in the crack, can be removed when filing the surfaces smooth. Soldering is also useful in bushing a pivot hole that is too close to the edge of a plate, causing a breakout to occur.

Bushings for wooden clocks are dealt with in Step 28 of the American 30-hour weight wooden movement (see page 52).

Bushings: *Making Your Own.* Aside from the large bushings dealt with under the two previous headings, it is best to make bushings from the stock brass tubes known as "bushing wires." These come in assorted, apparently arbitrary diameters. Reordering will not necessarily result in a resupply of tubes of the same diameter. However, this does not pose a problem. Typical diameters, measured in thousandths of an inch, are these: 70, 72, 78, 83, 88, 93, 99, 104, 108, 115, 136, and 138.

Most of the more common clocks have movements with plates 0.060"–0.065" thick. Bushings suitable for these plates will therefore be some 0.075" long, a length that allows a few thousandths of an inch at each end for riveting and facing off. Pivots vary in size, and typical diameters of bushings mostly range from 0.075" to 0.120". The 0.125"-thick plates of longcase clocks obviously require bushings longer than 0.075", so a few bushings 0.135" long, cut from a few of the larger diameter tubes, will also be useful. Naturally; fewer bushings of this type will be required.

Setting up for any job always takes time. It is therefore very shortsighted to make only sufficient units to fulfill immediate needs. This is particularly so when future requirements are a certainty. With this idea in mind, it makes sense to cut up a whole 3"-tube into bushings, rather than make, say, six and need to set up again a week later to make a few more.

Cutting bushings can be done quite rapidly on the lathe. In order to prevent them from falling into the swarf, or rolling off the lathe bed, after they have been cut, use the setup shown in Fig. 51. First grind a narrow parting tool, as shown, to a width of 0.015" or a little less. (A width of 0.010" will save more brass, but, because a tool of this width is more easily broken, we shall assume a width of 0.015".)

Turn the topslide of the lathe so that it is parallel to the bed, and set the lathe for about 800 RPM. Mount the tube in the headstock chuck (or collet), leaving about ¾" projecting. Take about 5" of 0.010"-diameter steel music wire, and make a fishhook bend about ⅜" long at one end of it. Set this end centrally in the tailstock chuck, with the bent section between two of the jaws. Close the chuck gently by hand until it just grips the wire hard enough to keep it from falling out. Do not, however, use the key to tighten the chuck, as this could strain its jaws.

Run the tailstock up, and insert an inch or more of the wire into the tube, leaving about 3½" still visible.

Move the topslide to its right-hand limit, and then turn it slowly back again until the micrometer scale reads 0, taking care not to overshoot. The topslide is now ready to move to the left—without backlash. In the interests of accuracy, this method must always be used when using screw feeds, as backlash is ever present. For travel to the right, the procedure would be reversed.

Moving the main carriage and the cross-slide, bring the parting tool up to the position shown in Fig. 51. Lock the carriage to the bed of the lathe. Using the cross-slide, withdraw the tool until it is clear of the tube, and then switch the lathe on. Now move the tool 0.090″ to the left, measured on the topslide scale. (This figure is the sum of 0.075″ [bushing length] and 0.015″ [cutter width].) Now run the tool in with the cross-slide and part off a bushing. This will then be retained on the wire, as shown in Fig. 51. Repeat this operation until only a stub of tube is left; this stub will be useful when the odd long bushing is required. Do not leave more than about ¾″ of tube projecting from the collet or chatter may occur. The thinner the tube, the less the projection.

When all the bushings are on the wire, slide the tailstock away and bend up the other end of the wire. Cut off the excess and store the bushings on the wire for future use. If the end of the wire is bent sharply at just the right angle, the bushings will stay on the wire yet still be removable with a firm pull. I find this a very convenient system of cutting and storing bushings. A small spur will be observed on each bushing; this is left when the bushing parts off from the main tube. When bushing a plate, always put this spur away from the punch; otherwise it will interfere with the punch and prevent a true blow.

Buying a Clock: There are so many aspects to consider in buying a clock that it is almost impossible to rank them in order of importance. Among the essentials is having a good idea of current prices; otherwise you could pay too much. You must also examine the clock closely to see that it is indeed what it appears to be; a casual glance is not enough. The presence of hands does not necessarily mean that the movement behind the dial is of the same make as the case, or is complete, or is there at all. A disproportionately short or long pendulum indicates a wrong movement. Do not buy a clock if you are forbidden to examine the movement fully and in a good light. New screw holes do not always mean that the movement has been changed, however; the feet may simply have been rotated because the screws no longer held.

If a grave shortage of parts is discovered and you think you can replace them, point the matter out to the owner, and offer a much-reduced price. Most clocks require overhauling, so take this into account, but do not reject a clock on these grounds alone if it is otherwise to your liking. Only experience will tell you what movements to expect in what cases; there are perfectly authentic alternatives in some instances.

Should the case be painted, as some are, the wood or veneer underneath may be anything from pine to surprisingly beautiful rosewood. It is important to know what to expect, so examine the case for clues of its presence. Paint can of course be stripped off, with the underlying wood or veneer generally unaffected by it; all the same, the sight of paint on such beautiful wood surfaces grieves the spirit. Very thick paint will sometimes chip off. Once, having filed the edge of such a chip at an angle, I counted twenty-seven separate coatings of different colors; and the total thickness was about 0.120″.

Restoring a solid-wood clock case is not too difficult, and is quite a popular pastime. Repainting a dial correctly, however, is not as easy as it may appear, and it is important to keep to the original design. A poor dial, therefore, affects the price—downwards. A little his-

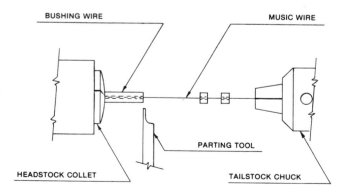

Fig. 51. Making Bushings

torical knowledge comes in handy here, for example in spotting an English longcase "marriage" (with the dial too young for the case or vice versa).

Most members of the public do not know the type—or even the proper name—of the clock that they are advertising for sale, and they can be singularly inarticulate in describing it over the telephone. If it sounds interesting, go see it at once.

One particular difficulty which you will experience is when an owner has had the clock valued by a friend or dealer, for he will then stick pretty rigidly to the quoted price, whether it is realistic or not. You can only point out that said friend certainly did not buy the clock at his valuation.

Dealers usually know a little about a lot; they cannot specialize in everything they sell. A specialist, of course, knows a lot about a little. It will not be too long before you yourself become a specialist who knows more about clocks than most dealers do. Most people are tolerably straightforward to deal with; misrepresentation, along with its attendant difficulties, is due more often to honest misinformation than to a desire to deceive. Still, the consequent difficulties can be serious: As an example, I once heard a dealer try to sell a "rosewood" mantel clock to a customer—at a high price—when I could plainly see that it was finished with adamantine, a celluloid-like material. It is interesting to speculate on what would have happened had the customer bought the clock and then tried to refinish it!

Before doing too much buying, then, go around the antique shops, junk shops, and flea markets until you have some idea of prices. Always remember that you are seeing asking prices; selling prices are always lower. It is not wise to blindly believe what the seller tells you—even though he may believe it himself. The pendulum of a "thoroughly overhauled" clock, for example, will not swing in a weary arc ¼" long. Also, reproductions are increasing in number; make sure you do not buy one by mistake.

People do strange things to their clocks. I have found suspension springs made from masking tape, cotton thread, a razor blade, and a strip of film. Inside clocks I have found false teeth, toothpicks, safety pins, thumbtacks, bobby pins, and the nests of mice and mud-dauber wasps. You will no doubt be able to augment this list as you enhance your collection.

One last word of advice: If you find a clock that is to your liking, and the price is reasonably fair—buy it right then and there. It is a curious fact that if, after only a short delay, you decide to buy, the clock is almost certain to be gone when you go back for it.

Calendars: The mechanical calendars found on clocks range from the simple version found on the English longcase clock to the perpetual calendar of double-dial American clocks.

The arrangement most commonly found in the English longcase is not too satisfactory. Basically a pin on the hour gear engages twice a day with a sixty-two-tooth gear. The latter gear is then engraved or painted to show the date through an aperture, or has a hand fastened to its arbor to indicate the date on a small dial or "calendar bit," as it is called. Although this is all perfectly sound in concept, in practice the pin cannot be placed quite close enough to the center of rotation to achieve a satisfactory mesh with the gear, owing to the relatively large diameter of the hour arbor. In this situation, the pin may butt against the tip of a tooth and become bent, or it may turn the gear two teeth at a time.

A far more reliable system is also found in the same type of clock, when 2:1 reduction gearing is used to obtain a gear that turns on a separate arbor, once a day. As here, the arbor is small, and the calendar gear needs only thirty-one teeth, a satisfactory mesh is readily obtained. The same system is also used on the successful simple calendars of American clocks. As the dials of these simple calendars are based on thirty-one positions, they must be reset by hand on the first day of any month of less than thirty-one days. However, this is a small price to pay for reliability.

More complicated calendar mechanisms by inventors such as Galusha Maranville (ca. 1861) and Daniel Jackson Gale (ca. 1869) are to be found, but they are not by any means common. Gale's clocks are quite rare, and much sought after.

Double-dial calendar clocks, when found, most often contain mechanisms invented by Benjamin B. Lewis (ca. 1862) and Henry Bishop Horton (ca. 1865). The Ithaca Clock Co. and Seth Thomas Clock Co. are the two best-known firms normally associated with double-dial calendar clocks. The roll-type calendar mechanisms used in them are quite similar in design.

In contrast to the relatively large rollers used in the above-mentioned clocks to denote the day of the week and month of the year, Lewis's elegant mechanism is much smaller and neater. However, the roller mechanism is a self-contained unit which displays all the calendar information on one dial. In Lewis's scheme, the clock movement itself must be modified, so that the day of the week can be indicated on the usual clock dial; the month and day of the month are shown on the lower dial.

Both the roller mechanism and that of Lewis are the perpetual type, which takes leap years into account. Although a good deal of thought and ingenuity went into the design of such calendar mechanisms, I do not consider them to be absolutely reliable over long periods. I usually know what day of the week it happens to be, although the day of the month often eludes me. To aid me in this area I prefer a simple calendar to a perpetual one, while not forgetting that without such devices horology would be much the poorer.

Cannon pinion: the pinion on the minute arbor—the first of the 12:1 reduction gears by which the hour arbor is driven from the minute arbor.

Center gear: the gear on the minute arbor. In English longcase clocks it is part of the actual train. In most American clocks it is separate, being driven from a gear of the train. (Removing a gear of the actual train stops the mechanism from running; removing a gear *driven from* the train does not stop it.)

Chain: Most clock chains are either of the ladder type (see page 38), or made up of simple oval links. Since the chains must fit the teeth of sprockets or toothed pulleys, the size of each link is critical, as is the gauge of wire from which it is formed.

Chainmaker: While it is obviously preferable to buy, rather than make, replacement chains for old 30-hour longcase clocks, the exact chain is not always available. This is due to the difficulty of duplicating each of the wide variety of chains used by the original makers. (For a practical analysis of the problems involved, see under **Huygens system.**) For this reason I designed and built the manual chainmaker of Fig. 52.

Because it takes several hours to make the eleven feet or so of chain required for such a clock, this machine is not an ideal long-term solution to the problem. It will, however, turn out uniform, well-shaped links, and so can create a custom-made chain for a cherished old-timer.

The chainmaker is made of cold-rolled mild steel, tool steel, and a little brass. It took a few hours to design and about twelve hours to build. The proportions of Fig. 52 are fairly accurate, so scaling is possible; because chain links differ considerably, exact dimensions are not applicable. Given the general idea, if you build such a device you will no doubt wish to incorporate your own modifications.

The circular base shown in Fig. 52a was used simply because it happened to be available; a square base is also acceptable. The stop bar is retained at each end by a no. 8 screw and spacer and by two stop pins. The stop pins

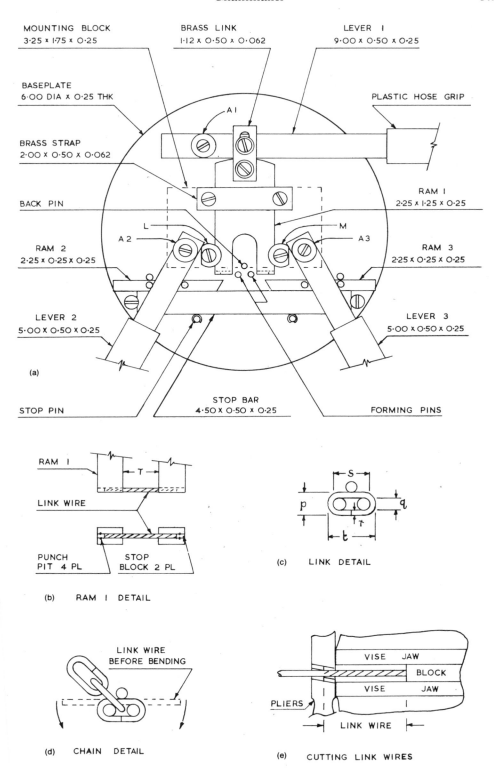

Fig. 52. Chainmaker

have sleeves over them, the position of the stop bar being determined by the wall thickness of the sleeve. Adjustable screws, running in blocks, make a good alternative to stop pins. The exact position of the stop bar, the throat width T (Fig. 52b), the diameter and positions of the forming pins (52a), and the length of the retaining slots, are all determined by the dimensions of the desired chain link. It is essential to know exactly the number of links per foot and the other link measurements before attempting to make any of the above critical parts.

A few links from different chains are very useful when matching a chain to a spiked pulley. (However, see under **Huygens system** before starting to build, for more aspects are involved than appear at first glance.) Some initial experimenting is necessary before deciding upon a given link. This involves varying the spacing of the chainmaker's forming pins, using a piece of scrap steel, until the correct dimensions are obtained (see Fig. 52c). Dimension t (Fig. 52c) determines T, the width of the throat, which will be a few thousandths of an inch greater than t, to allow clearance during forming.

The links are made from mild-steel wire, typically 0.063″ in diameter. It is available in hardware stores in coils of fifty feet or more, and is plated with cadmium or similar plating. If a choice is offered, take the stiffer wire for preference. This wire can be used to make links for most of the required chains.

The wire is straightened by holding one end in the vise and pulling two closely held pieces of wood along it, using a firm grip. The straightened wire is then cut to exactly the right length to form the desired link. In the case mentioned earlier the requirement was for thirty-six links per foot; the length of wire to form one link works out as $^{31}/_{32}$″. I must stress that in circumstances such as these, $^{15}/_{16}$″ or 1″ will *not* do; the length must be $^{31}/_{32}$″. If the link is too long, it will be difficult to close, and may jam the rams; if too short, a gap will result.

Fig. 52c shows a correctly formed link. Dimension p is the outside width of the link; it is the distance between the sides of the pulley minus a working clearance. Dimension q is the diameter of the forming pin; it is the thickness of the pulley spikes plus a working clearance. The wire thickness r therefore cannot exceed ½ (p−q). Dimension s is the inside length of the link and is determined by the spike-spacing on the pulley, which also dictates the number of links per foot. The distance between the centers of the forming pins is therefore s−q. The back pin is located on the vertical center line, and is spaced from the two forming pins by the distance r plus a few thousandths of an inch clearance, for the link wire must be inserted into this gap. If the back pin were not present, the link wire would bow during forming, whereas the sides of the link should be almost straight. The position of the back pin is therefore critical.

When the length of the link has been determined, it becomes necessary to find a reasonably easy way of cutting off about four hundred lengths of wire, to a tolerance of ±0.005″. Fig. 52e shows a very simple way of doing this. Clamp a block of steel about ½″ thick in the vise, and position a pair of heavy pliers alongside the end of the vise jaws as shown. With the pliers open, set the wire into the pliers' cutting notches in the side of the joint, so that the end of the wire butts against the block. Making sure that the wire is parallel to the vise jaws, cut off a length and measure it with a micrometer. Adjust the block if necessary, repeating the process until the correct length is obtained. The ends will be reasonably square cut. As long as parallelism is maintained, this method is accurate enough for the intended purpose.

Push a piece of cardboard through the vise, so that it extends as a channel beneath the pliers, and set up a small box under the end of it. This simple arrangement will allow the cut link wires to drop into the box, ready for use in the chainmaker.

Fig. 52b shows the detail of ram no. 1 with a link wire in position. It is advisable not to set the two stop blocks into the slot until a test run has confirmed the length selected for the link wire.

Clamp the chainmaker in the vise by means of the mounting block screwed to the underside of the base plate. Push lever no. 1 away in the general direction of "twelve o'clock" and move levers 2 and 3 apart until they stop against the heads of the no. 8 screws which form end stops for the stop bar. These three levers swivel on arbors at A1, A2 and A3, lever no. 1 resting directly on the base plate and levers 2 and 3 on ¼" × ¼" brass collets at A2 and A3. (See Fig. 52a.) Levers 2 and 3 move rams 2 and 3 by means of a pair of pins set into the top of each ram. As the levers will be in use continuously for several hours, it is advisable to push a short length of plastic garden hose over the end of each one, to avoid sore fingers. Old drill stubs can be used throughout as pins.

Now load one of the link wires into the slot of ram 1 and center it. It should lie between the back pin and the two forming pins, as shown in Fig. 52d. This first link wire is a little difficult to load, but all the others will be much easier, because the last link formed is used to grip each succeeding link wire.

Now apply light pressure to lever no. 1 so that the link wire is held against the forming pins, and center the link wire in the slots of ram no. 1. The ends of the link wire should now be centered about the forming pins, also. Apply more pressure to lever no. 1, pulling it towards the "three o'clock" position. The ends of the link wire will now leave the slot and curve round the forming pins. If this proves difficult, it probably means that the throat of ram no. 1 is too narrow. (Note in Fig. 52 that the upper hole in the brass link connected to lever no. 1 is oval, so that lever no. 1 can press directly upon the rounded top of ram no. 1; the brass link is therefore used only to retract ram no. 1.)

The link wire should now be bent into an inverted U, with the ends of ram no. 1 resting against the ends of rams 2 and 3.

Ram no. 1 must slide as a lathe slide does—firmly, but not too freely. A little tension is applied by means of the brass strap and the two overhanging washers and collars mounted on the no. 8 screws on either side of the throat end, at L and M in Fig. 52a.

Leaving ram no. 1 in this position, bring levers 2 and 3 evenly together. This will cause rams 2 and 3 to push the ends of the link wire further round the forming pins, until the ends of the link wire butt together, thus making it into a closed link.

Now retract all the levers, and pry the link off the forming pins, using a small screwdriver. (After a little use, the pins will become very slightly tapered and the links will come off easily, assisted by the chain of those already made.) Now fit a new link wire through the link just made, and set it up in the slot of ram 1, as shown in Fig. 52d. It is easy to retain control over each successive link wire by holding the just-made link between thumb and finger so that the fresh link wire presses against the fingernails.

Repeat the above procedure, and the chain will begin to grow out of the machine in the direction of "eleven o'clock." Keep the chain on the bench at all times. When a chain of 15" or so is obtained, check the links for flaws; then check for twists by letting the chain hang. If the link wires have been cut consistently, and the links properly closed, the quality will be good. A slightly open link can be closed up in the pliers by pressing across its width; it is not as effective to press from end to end, unless the gap is wide. Note here, by the way, that it is best to open and close a chain link by twisting it lengthwise rather than levering one end outward.

Check the number of links per foot. Then try the chain on the spiked pulley. It should fit snugly and show no tendency to ride up when pulled round. If results indicate that a slightly shorter link is desirable, it is feasible to make, say, a thirty-six-links-per-foot chain into a thirty-seven-links-per-foot one without altering the chainmaker.

The method of doing this is to cut link wires 0.010″ shorter, and then to make the chain as before. Each link is then closed up in the pliers in the end-to-end mode. With a little experimenting, it is thus possible to make two slightly but significantly different chains, without major surgery on the chainmaker.

When the correct link-wire length has been ascertained, press the two small stop blocks into the slot of ram no. 1 (see Fig. 52b). Lightly swage a little metal over them with a center punch, if necessary. The link wires will then be centered automatically.

Now pull the finished chain through fine emery cloth to smooth it and get rid of any loose scale or powdering of the finish. The resulting chain should hang straight and be of an even, pleasingly burnished appearance.

When fitting the chain to the clock, take care not to include a built-in twist in it. Twists almost always develop during normal running and winding, but these are random and self-rectifying.

Improvements to the chainmaker might possibly include case-hardening the rams, and mounting the back pin and forming pins on an interchangeable plug. The latter, in conjunction with several no. 1 rams of slightly differing throat sizes, would enable the creation of a wider variety of links. An automatic machine would, of course, be even more interesting to design!

Chiming Clock: usually a clock with three trains: time, strike, and chime. Occasionally a two-train clock chimes as well as strikes—for example, a three-quarter Westminster, which chimes at 15, 30, and 45 minutes, but not at 60, when it simply strikes the hour. Unfortunately, many people say a clock chimes when in reality it strikes instead, and this is confusing. Chimes, unlike striking, occur at the quarters; they do not indicate which hour of the day it is.

Chops: small pieces of metal, usually brass, which lie on either side of a third piece; for example, the pieces at the ends of a suspension spring.

Cleaning machine: The clock enthusiast who requires a cleaning machine has very little choice available to him. He can buy an ultrasonic model, which costs several hundred dollars, or make some form of mechanically agitated device himself. Ultrasonic machines are efficient; however, their cost rules them out for most people, so a home-built machine appears to have considerable merit.

The need for some form of machine soon becomes apparent to anyone who tries simply immersing clock parts in cleaning solvent and leaving them to soak for half an hour. As there is no agitation, the dirt on the parts is attacked only by the liquid that touches it; as this becomes exhausted, a partial shield forms between the part and the surrounding unused liquid. The result is that only partial cleaning occurs, and a rather thick, viscous coating develops on the parts.

The most obvious solution is to use an impeller to swirl the liquid around, but this has a drawback in that the resultant unidirectional flow deposits fibers on certain sharp points, such as the teeth of the escape gear. Also, because velocities are unequally distributed, some parts are cleaned better than others; hence, a random reversing flow would seem to be indicated.

Whatever the method of cleaning, a rinse is necessary afterwards. It is easiest to use a basket of mesh wire or perforated metal for easy transference. A mesh basket made of copper fly screening is feasible, and is shown in Fig. 53. Any form of mesh, however, tends to collect fibers at wire crossing points, and these are troublesome to remove. The perforated metal basket of Fig. 54 does not have this defect and is easy to clean. As the holes in it are large enough to pass small nuts and screws, however, a small mesh basket is necessary to contain them. This combination works well for larger clocks, while a copper or brass mesh basket is fine for small ones.

An easy way of obtaining the desired reversing flow is to keep the parts fully immersed while alternately raising and lowering them. This action continually brings fresh liquid to attack the dirt on the clock parts, and so cleans them evenly while preventing the formation of a coating. A travel of 2″, repeated from fifteen to thirty times a minute, provides good results. The time required to thoroughly clean the parts varies with the strength of the cleaning solvent. When the solvent is fresh, twenty minutes is usually sufficient.

Certain pieces never seem to attain a bright clean finish, even after long immersion in fresh solvent. They may be clean, but the brass remains dull and perhaps pitted; only vigorous hand-polishing with fine emery and abrasive wool produces any kind of shine at all. This effect is presumably due to the type of brass and its environment. Since such parts are frequently eroded by verdigris, and are cracked and friable, they must be cleaned with care.

The required rise and fall for the machine is best obtained, piston fashion, from the rotary action of a small gear motor. Such motors are to be found in discarded appliances such as humidifiers and barbecues, and in surplus stores and scrap yards. The output shaft is usually ¼″, ⁵⁄₁₆″, or ⅜″ in diameter, all of which are very suitable; the direction of rotation is not important.

In designing a cleaning machine, it is essential to make provision for the largest clock movement likely to come to hand. For most people, this is perhaps the movement of a longcase or an old punch clock. Although 1 imperial gallon of solvent will completely cover such a movement, it is not sufficient to cover a basket as well, particularly when allowance must be made for vertical motion.

Fig. 53 shows a wire mesh basket suitable for use with a small 11″-high metal wastebasket. This is a useful size when only a gallon of solvent is available. Although a parallel-sided container is desirable, most wastebaskets taper; the smaller versions typically have diameters of about 8″ at the bottom and 11″ at the top,

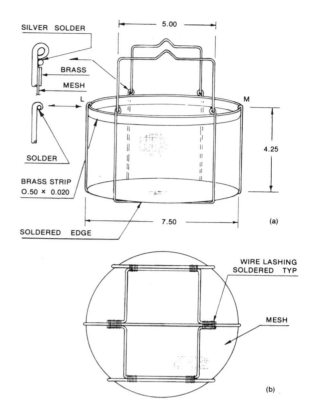

Fig. 53. Wire Mesh Basket

which allows them to hold about 2.8 gallons. In a container like this, a gallon of solvent will have a depth of some 4¾″. This is quite adequate for most purposes, as the basket of Fig. 53 will hold the great majority of commonly found movements. It may possibly accept a 30-hour longcase movement. Note, however, that the plates of the 8-day version, if stood on edge, may not be completely covered by solvent, and a further half gallon becomes necessary.

The designs shown in Figs. 53 and 54 are intended to present general constructional ideas, rather than precise layouts. The dimensions given are typical; they can easily be changed to suit the reader's own container.

The wire mesh basket is made of copper fly screening, soft soldered together. The supporting frames are made of thoroughly cleaned coathanger wire, the joints being lashed with

0.020"-diameter tinned copper wire and then soft soldered. Before bending the coathanger wire, scrupulously clean it along its entire length with emery cloth. This ensures good soldered joints and also prevents the pollution of the cleaning solvent with varnish or paint dissolved from the coathangers. For hooking the basket on, the rectangular handles shown are much more convenient to hold than the rather more obvious triangular ones which might initially suggest themselves.

The side is simply soldered to the bottom all around (Fig. 53). The top of the side is soldered to a strip of brass which rests just under the circular rim wire to which the upright members are silver soldered. In order to make a neater joint, the ends of the two uprights that do not secure the handles are filed flat before being bent over the rim wire. These two joints are shown in Fig. 53a at points L and M.

The copper mesh itself is not soldered to the wire frame, but merely rests in it. It cannot slip out, however, because the brass strip around its top sits underneath the rim wire. Although it is preferable to make the side in one piece, two or three pieces joined by soldered vertical seams are also acceptable.

Fig. 54a shows a perforated brass basket which fits well into a 16"-high, 5-gallon container of some 10½" in diameter. Two gallons of solvent here give a depth of about 6.4", and since the 9"-diameter brass basket will accept the plates of an 8-day longcase movement laid flat, all clock parts are well immersed throughout the cycle.

Small parts—such as the gathering pallet, screws, nuts, washers—must be put in the separate auxiliary basket (Fig. 54b). This is a roughly oval piece of copper fly screening with a strip of 0.010" brass soldered around it; the seam in the strip is also soft soldered.

As perforated brass is difficult to find, it is probably best not to waste too much time looking for it, and to simply cut out the required pieces from 0.020" brass shim stock and drill holes in them.

The circular bottom has twelve small, bent-up tabs equispaced around the edge. These are drilled with a no. 32 drill to take no. 4-40 NF × 3/16" screws and thereby secure the bottom to the side. The side can be conveniently made up of four pieces, each 7½" long, which overlap each other by about ⅜". The bottom screw of each of the four overlaps passes through a tab, and the bottom is soldered to the side between tabs. The other two screws that secure the overlaps are countersunk into four strips of 0.050" brass, 2⅜" long × ½" wide, which form the handle lugs. These lugs are twisted at the top, so that the handles fall tidily into the basket for storage.

The rim of the basket is strengthened by a ring of ⅛"-diameter brass wire, which is soldered in place. The joint, or joints, in the ring are of the scarf type, as shown in Fig. 54c. A ⅜"-long piece of thin-wall brass tubing covers the joint and gives it considerable strength when all is soldered together. Being short, the tube is easily turned up on a lathe, the wall thickness being about 0.020".

In the center of each quarter-side, drill two

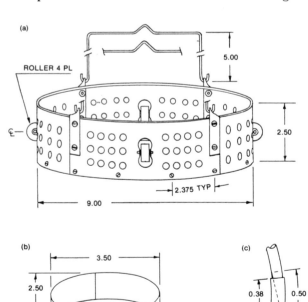

Fig. 54. Perforated Metal Basket

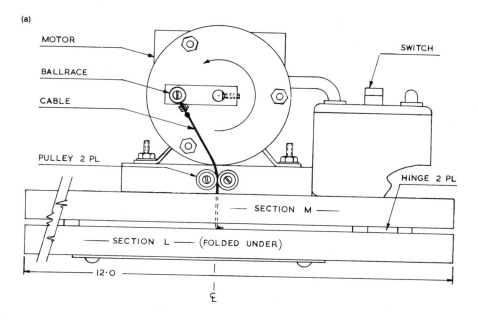

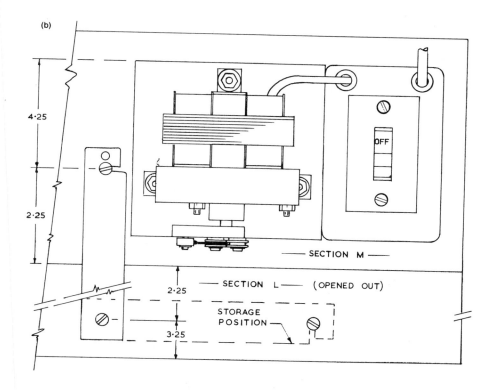

Fig. 55. Cleaning Machine

½"-diameter holes vertically spaced ⅝" apart. By cutting vertically through the metal between them and bending the resulting edges outward, you can then form two small lugs. Drill a 0.100"-diameter hole in each lug, before bending; these two holes can be used to support a short shaft to carry a small roller between the lugs, as shown in Fig. 54a.

There are four such rollers, ⅜" wide and parted off from ¾"-diameter nylon rod. They should have a washer on each side of them on the shaft, and are optionally fitted with brass bushings; the ends of the shaft are soldered to the lugs. The rollers make for quiet running, and prevent scoring and wear on the basket. A plastic container would become very rough in a short time if rollers were not fitted, and a steel one would be quite noisy as it scraped the basket.

All vent holes are ¼" in diameter; there are 96 of them in the side, 24 in each quarter-side. There are 157 holes in the bottom, laid out in a pattern of concentric circles of radii ½", 1", 1½", and so on up to 4". There is a hole opposite each tab, and two more holes between them on the outermost circle, the overall spacing being similar to that of the side—that is, ⅝" between centers. All screws and nuts should preferably be of brass or stainless steel. This need not apply to handles, however; I find that the mild-steel coathanger wire handles show no signs of rusting.

Mark out all the vent holes, and center-punch and drill them, removing all burrs before assembly. Do not be tempted to drill random holes to save time. A basket made as described above will last almost indefinitely, assuming reasonable usage, so it is best made in a workmanlike manner.

Fig. 55 shows a typical, small gear-motor

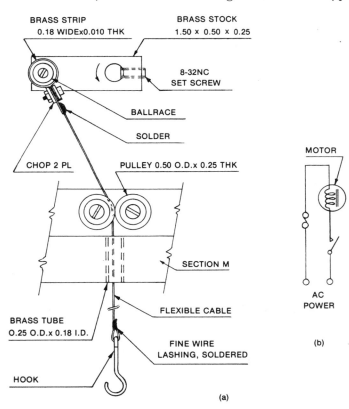

Fig. 56. Drive Detail

mounted on a 12″ × 12″ square of ¾″-thick board. For portability and ease of storage the mounting board is hinged, so that section L can fold under section M. The folded machine then fits endways into the mesh basket of Fig. 53, which, with its handles raised, sits in the 2.8-gallon wastebasket, to which a pail-type carrying handle is easily fitted. When the unit is operational the metal strap shown in Fig. 56b prevents accidental folding. The strap size is not critical, but it is best made of steel, 6″ × 1¼″ × 0.035″ being a good size; the strapping used on large crates is ideal. The slotted free end fits under the head of a round-headed screw in both positions, to hold it in place.

Section M has to be larger than section L, as the cable must drop centrally. A small brass tube is useful here to guide it (see Fig. 56). Four ¼″-wooden dowels, ¼″ high, locate the board on its container. They also serve to keep the two sections parallel when folded, and thus provide a space for the cable and hook; they are positioned so as to miss each other in the folded position.

The drive detail is given in Fig. 56a. A small ball race is mounted on the end of the driving arm, and around it is clamped a narrow strip of 0.015″ brass to which the flexible cable is soldered, after passing through a hole. A fishing leader makes a good cable; the plastic coating need not be removed, except for soldering. The cable must be cut to a length such that, when the crank is at "six o'clock," the basket is about ¾″ above the bottom of the container. As the soldered joints do not flex, they cause no trouble, and the cable will have a long life.

The two pulleys almost touch each other, so the cable cannot slip out. The larger the pulleys' diameter, the kinder they are to the cable; hence, 1″-diameter pulleys would be an improvement on those of Fig. 56, though a thicker motor base may be required to mount them.

The circuit diagram is simple enough, as shown in Fig. 56b; the fuse should be of the minimum amperage that will not blow under normal running conditions.

When considering container volumes, the following data are useful:

$$v_{cyl.} = \pi r^2 h$$

where $v_{cyl.}$ = cylinder volume in cubic inches
$\pi \cong 3.14$
r = radius in inches
h = height in inches

U.S. gallons = imperial gallons × 1.2

cubic inches = imperial gallons × 277.27
= U.S. gallons × 231

Cleaning process: Several good concentrated solvents are available from suppliers. With the addition of water, a pint of concentrate provides a gallon of cleaning solution. Other concentrates are available from oil companies, in 5-gallon containers, and these may be a better buy. As all such solutions may well prove toxic, they must be kept in closed containers and used with care. Special rinses are even more expensive than the concentrates, so it is best to use a water-soluble concentrate which can be swilled off with hot water, followed by a further rinse in white gas (camp-stove fuel).

White gas will remove any residual water and leave the parts clean and dry; it is a relatively cheap and effective final rinse. As it is essentially refined gasoline, it must be treated as such: Winter or summer, never bring it into a house or basement; always work outside in a well-ventilated area, and in the total absence of flame or sparks of any kind. White gas vaporizes readily, and if used indoors, furnace flames or a spark from a contactor can easily cause explosion and fire.

To use the cleaning machine described previously, put the clock parts in the basket and give them from twenty to thirty minutes of wash-cycling. Remove the basket, swill the parts with hot water, and then immerse them fully or partially in the white gas. Brush the parts individually as you remove them, and dry them off quickly. (A heat lamp is a good way.)

If desired, polishing the clock plates may now be done; all polish residues must be thoroughly removed or verdigris will form in due course.

If solvent is to be permanently stored in the cleaning machine, a seal must be installed around the top to avoid the escape of vapors which may be unpleasant or toxic.

Clicks: The term *click* is a very descriptive one, for that is just what this part does. However, only clockmen seem to use it; others use *pawl*. Click or pawl, it rests lightly on its partner, a ratchet gear, and a one-way rotation results.

In an ogee movement, the pressure on the clicks is small, but in most other clocks, particularly the English fusee, the stress is much greater. In most cases click failure will result in damage to the clock or to the fingers of the person winding it, or to both. Weight-driven clocks are crank wound, and are thus easier to control when a click fails during winding. It is important to check the clicks during overhaul and ensure their correct operation.

When a click is loose, it tends to jump out of the ratchet-gear teeth slightly axially, instead of meshing with them properly in the radial plane. It then becomes essential to tighten up the rivet on which the click turns. To do this, lay the no. 1 gear, click downwards, on an anvil, and carefully hammer the rivet tight again, testing the click periodically for freedom. If the click "seizes up," as it often does, it can easily be freed again: Turn the gear over and rest it on the anvil again, click upwards. Now set a hollow punch *over* the head of the rivet, so that it rests on the click itself, and deliver a light tap. Repeat as necessary, and the click will loosen. Check that the rivet is still firm.

When there is a zinc or brass disc on the arbor, drill a hole in the disc to enable punch access to tighten the rivet. Use the same flat punch, inverted, as an anvil, if it becomes necessary to free the click with a hollow punch. This operation can be carried out through the clearance hole in the zinc disc, without removing the mainspring, should one be present.

For English longcase clicks, see page 75.

Clock stand: Fig. 57 shows how a versatile clock stand can be made from a retort stand. However, a cast-iron base from an old floor lamp and a 16″ length of ½″-diameter steel rod will serve equally well. This stand will accept almost all mantel and shelf clocks; wall clocks, in contrast, must be cased or hung on a wall or board for testing. To ensure rigidity, any clock movement in the stand must be mounted as low as possible. The stand may also be screwed down.

Two standard blocks are used; the only items to be made are the holding arms. These are simply two lengths of mild-steel rod of the dimensions shown in Fig. 57c. The two sets of V-grooves are filed out to suit, using square and triangular files. The blocks lock to the rods with Allen cap screws. The larger type of retort-stand clamp, which has convenient thumbscrews, may also be used.

It is a good idea to push a piece of stiff cardboard onto the vertical rod, so that it hangs just above the clock movement to keep off dust.

Collet: a collar or flange that usually carries a gear, cam, lifter, or anchor.

Count gear: a gear which incorporates a count wheel by means of extra-deep intertooth slots spaced apart varying degrees—one tooth, two teeth, three teeth, etc.—to count the hours of the day up to twelve. Its teeth enable it to be directly driven. (See Figs 2 and 7d.)

Count wheel: a wheel with its periphery divided by slots into sections proportional in length to the hour of the day—that is, one

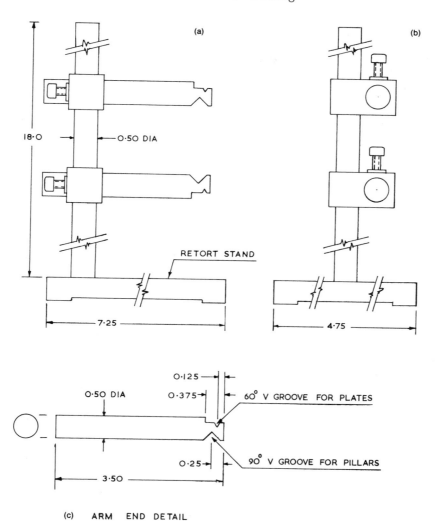

Fig. 57. Clock Stand

hour, two hours, three hours, and so on up to twelve. A count wheel cannot be directly driven; it must be fixed to a gear or arbor.

Count hook: a lifter arm associated with a count gear or count wheel. (See Fig. 7.)

Crutch: the linkage between the anchor and the pendulum.

Dapping: the process of using dapping blocks, cutters, and punches to make convex-concave discs, such as hand-retaining washers.

Drilling: A full treatment of this subject is beyond the scope of this book, but certain aspects deserve mention. Most of the holes drilled in clock repairing are of critical importance, and must be drilled accordingly. See page 134, under **Bushing:** *Plates,* for details on how to drill a hole of a specific size.

Most of the holes drilled in clock repair are in brass, and drills used for brass should be sharpened a little differently from those used for steel. As most people only possess one set of drills at best, little heed is paid to sharpening them for specific tasks and they are generally

Table 3. Drill Sizes

Drill Size No.	Decimal	Drill Size No.	Decimal	Drill Size No.	Decimal	Drill Size No.	Decimal
80	.0135	43	.0890	8	.1990	25/64	.3906
79	.0145	42	.0935	7	.2010	X	.3970
1/64	.0156	3/32	.0937	13/64	.2031	Y	.4040
78	.0160	41	.0960	6	.2040	13/32	.4062
77	.0180	40	.0980	5	.2055	Z	.4130
76	.0200	39	.0995	4	.2090	27/64	.4219
75	.0210	38	.1015	3	.2130	7/16	.4375
74	.0225	37	.1040	7/32	.2187	29/64	.4531
73	.0240	36	.1065	2	.2210	15/32	.4687
72	.0250	7/64	.1093	1	.2280	31/64	.4843
71	.0260	35	.1100	A	.2340	1/2	.5000
70	.0280	34	.1110	15/64	.2344	33/64	.5156
69	.0292	33	.1130	B	.2380	17/32	.5313
68	.0310	32	.1160	C	.2420	35/64	.5469
1/32	.0313	31	.1200	D	.2460	9/16	.5625
67	.0320	1/8	.1250	E 1/4	.2500	37/64	.5781
66	.0330	30	.1285	F	.2570	19/32	.5937
65	.0350	29	.1360	G	.2610	39/64	.6094
64	.0360	28	.1405	17/64	.2656	5/8	.6250
63	.0370	9/64	.1406	H	.2660	41/64	.6406
62	.0380	27	.1440	I	.2720	21/32	.6562
61	.0390	26	.1470	J	.2770	43/64	.6719
60	.0400	25	.1495	K	.2811	11/64	.6875
59	.0410	24	.1520	9/32	.2812	45/64	.7031
58	.0420	23	.1540	L	.2900	23/32	.7187
57	.0430	5/32	.1562	M	.2950	47/64	.7344
56	.0465	22	.1570	19/64	.2968	3/4	.7500
3/64	.0469	21	.1590	N	.3020	49/64	.7656
55	.0520	20	.1610	5/16	.3125	25/32	.7812
54	.0550	19	.1660	O	.3160	51/64	.7969
53	.0595	18	.1695	P	.3230	13/16	.8125
1/16	.0625	11/64	.1719	21/64	.3281	53/64	.8281
52	.0635	17	.1730	Q	.3320	27/32	.8437
51	.0670	16	.1770	R	.3390	55/64	.8594
50	.0700	15	.1800	11/32	.3437	7/8	.8750
49	.0730	14	.1820	S	.3480	57/64	.8906
48	.0760	13	.1850	T	.3580	29/32	.9062
5/64	.0781	3/16	.1875	23/64	.3594	59/64	.9219
47	.0785	12	.1890	U	.3680	15/16	.9375
46	.0810	11	.1910	3/8	.3750	61/64	.9531
45	.0820	10	.1935	V	.3770	31/32	.9687
44	.0860	9	.1960	W	.3860	63/64	.9844
						1	1.0000

Drilling

Table 4. Tapping Drill Sizes

Screw Size			Tap Drills		Clearance Hole Drills				Countersink 80–82° Dia.
					Close Fit		Free Fit		
No. or Dia.	Decimal Equivalent	Threads per inch	Drill Size	Dec. Equiv.	Drill Size	Dec. Equ.	Drill Size	Dec. Equ.	
0	.060	80	3/64	.0469	52	.0635	50	.0700	.119
1	.073	64	53	.0595	48	.0760	46	.0810	.146
		72	53	.0595					
2	.086	56	50	.0700	43	.0890	41	.0960	.172
		64	50	.0700					
3	.099	48	47	.0785	37	.1040	35	.1100	.199
		56	45	.0820					
4	.112	36	44	.0860	32	.1160	30	.1285	.225
		40	43	.0890					
		48	42	.0935					
5	.125	40	38	.1015	30	.1285	29	.1360	.252
		44	37	.1040					
6	.138	32	36	.1065	27	.1440	25	.1495	.279
		40	33	.1130					
8	.164	32	29	.1360	18	.1695	16	.1770	.332
		36	29	.1360					
10	.190	24	25	.1495	9	.1960	7	.2010	.385
		32	21	.1590					
12	.216	24	16	.1770	2	.2210	1	.2280	.438
		28	14	.1820					
14	.242	20	10	.1935	D	.2460	F	.2570	
		24	7	.2010					
1/4	.250	20	7	.2010	F	.2570	H	.2660	.507
		28	3	.2130					
5/16	.3125	18	F	.2570	P	.3230	Q	.3320	.635
		24	I	.2720					
3/8	.375	16	5/16	.3125	W	.3860	X	.3970	.762
		24	Q	.3320					
7/16	.4375	14	U	.3680	29/64	.4531	15/32	.4687	.812
		20	25/64	.3906					
1/2	.500	13	27/64	.4219	33/64	.5156	17/32	.5312	.875
		20	29/64	.4531					

Glossary 153

used as is for all metals. It thus frequently happens that a drill grabs the clock plate and jerks it upwards, which results in a hole sadly out of round.

However the drill is sharpened, it is unwise to drill holes much over 1/8" without clamping the clock plate down securely in the drill vise. In addition, for holes over 1/4", the drill vise should be clamped to the drill-press table. These two sizes are only approximations, and the condition of the drills must also be considered, but the principles still remain: Keep the drills sharp, and clamp the work down.

Other points worthy of mention are that the larger the drill, the slower the speed; and to never use an electric hand drill, for it will tear up the drill shanks until accuracy becomes impossible. Certain small holes are more conveniently drilled in a vise, in which case use a geared hand-brace. Although somewhat despised, this tool is easy to control, quite accurate when used correctly, and kind to drill shanks. As the thrust is on the center line, rather than below it as with an electric hand drill, it is easy to drift a hole to its correct position by angling the brace when starting to drill and then uprighting it when the hole is correctly placed.

Most holes undoubtedly should be drilled on a drill press, as accuracy is important. When not using the drill vise, always drill on a piece of flat wood; you will thus avoid the necklace of drill pits which disgraces so many drill-press tables.

Imperial "system" drills are a conglomeration of three separate series: fractions, letters, and numbers; these interlock on a stopgap basis. They are given in full, with decimal equivalents, in Table 3. Table 4 gives tapping data, which is useful in various clock-oriented projects.

Escape cock: The external support for the escape arbor.

Escapement:

1. *Recoil—Strip Pallet—American (external)*
2. *Recoil—Strip Pallet—American (internal), English and German*
3. *Recoil—Solid Pallet—American*
4. *Recoil—Solid Pallet—English Longcase*
5. *Recoil—Solid Pallet—English, French, and German*
6. *Deadbeat—Strip Pallet—American*
7. *Deadbeat—Solid Pallet—American*
8. *Deadbeat—Solid Pallet—English*
9. *Deadbeat—Solid Pallet—German*
10. *Deadbeat—Solid Pallet—Vienna Type*
11. *Brocot—Pin Pallet—American*
12. *Brocot—Pin Pallet—French*

All of the twelve escapement variations listed above are found in the clocks dealt with in this book; they are numbered here for ease of reference. Whether an anchor is of the strip or solid type is purely a matter of construction; the principle of operation is unaffected. The **verge escapement** is quite rare, and thus is only briefly reviewed, separately under that heading, to avoid any confusion of terms.

The escapement is undoubtedly the most critical part of a clock's mechanism, and it is essential that its action is thoroughly understood. Beginners sometimes ask why a clock is not made more powerful, so that it would run more reliably. This is done to some extent in tower clocks, where the relatively powerful main drive for the hands is not applied directly to the escapement. Here, a remontoire periodically raises a small weight, the fall of which drives the escapement with a constant force. Smaller clocks also sometimes use this system, but because it is somewhat involved and expensive, it is generally found only in master clocks, where accuracy is necessary. However, the basic fact is that only the minimum energy must emerge from the mechanism to maintain its governor, the pendulum,

Escapement

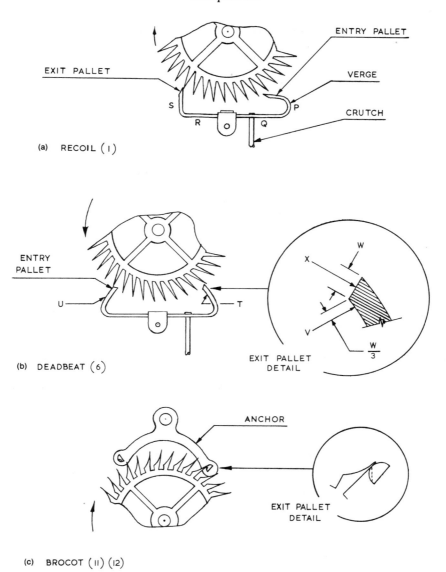

Fig. 58. Escapement Types

in business. If more energy were supplied, the pendulum would no longer be in command and erratic timekeeping would result. The function of the escapement is to supply this energy, in phase, to the pendulum to maintain its action, while influencing that action as little and as consistently as possible. (See also under **Pendulum.**)

Practically all escapements fall into two categories as listed above—i.e., recoil or deadbeat. The bounce of a recoil escapement is easily seen in the motion of the seconds hand of an

English longcase clock. The deadbeat escapement is distinguished by abrupt, recoilless action, clearly demonstrated by the steady, steplike motion of the seconds hand of a fine regulator. The recoil is a reliable, undemanding type of escapement for everyone to use, whereas the deadbeat is more delicate and more accurate.

Recoil no. 1 (Fig. 58a) is representative of the other recoil variants (2, 3, 4, and 5 in Fig. 59). Although most American strip-pallet verges are located outside the plates, internal placement is also common. Among English and German clocks, the verge is almost always riveted into a slot in an arbor, and so lies between the plates.

Another variant—not listed above, as it is seldom found—is the club-tooth deadbeat escapement. This variant has sharp-edged pallets which are actuated by relatively wide escape-gear teeth ending in sloping ramps. Welch, Spring and Co.'s "Verdi," a long-drop school clock, utilizes this escapement, and it performs well. Other interesting uncommon escapements include the pinwheel and the grasshopper.

In setting up an escapement, it should be borne in mind that the pallet tips must always embrace x-and-one-half teeth—never a whole number. Common values of x are 7 and 8, but other numbers also occur. If the half space were not included, the pallets could not enter the inter-tooth gaps, and the escapement would not function.

The way an escapement works is that a tooth presses against a pallet until it forces it away; this allows the escape gear to rotate until the other pallet stops another tooth spaced x-and-a-half tooth spaces from the first one. The action then repeats itself until the first pallet contacts again, and a rocking motion is thus set up which is transferred via the crutch to the pendulum to maintain its oscillation.

This simple explanation fulfills present requirements better than would an in-depth study of the various angles and distances involved in escapement design. Several authors have set forth their ideas on this intriguing subject, and the reader would be well advised to study their works and experiment with their findings; see under "Recommended Reading" in the back of this book.

The short periods when the escape gear is free of the pallets, and can therefore rotate briefly, are known as "drops." In a correctly set-up escapement, the drops are brief and of equal duration. Because the drops represent wasted energy, they must be reduced to the minimum necessary to provide clearance. A large drop means that instead of driving the pendulum the escape-gear teeth are simply hitting the pallets unduly hard, thereby digging undesirable pits in them.

Before attempting to make any adjustment to any verge or anchor, lightly run an old needle file across the backs of the pallets (points P and S in Fig. 58a), to check the hardness of the steel. If the file cuts, the steel is soft; if it slips, with a rather squeaky sound, the steel is hard. The verge may be soft all over, as are many no. 2 types, or in a few cases hard all over, but most will be hardened in the regions of P and S and soft between Q and R. This is important to remember, because soft steel can be bent, whereas hard steel cannot; it breaks instead. Many people break verges by trying to bend them in the hardened state. By using the above test, such mistakes can be avoided.

If it is necessary to adjust a hard-steel pallet, it must first be softened by being heated to red hot and then allowed to cool rather slowly, for, say, over thirty seconds. (See under **Hardening** for further information.) Take care not to heat the saddle or the crutch with the flame.

Although a change in any part of an escapement affects the whole, in practice the two drops may be considered as being individually adjustable. It is easy to verify that moving the verge closer to the escape gear reduces the

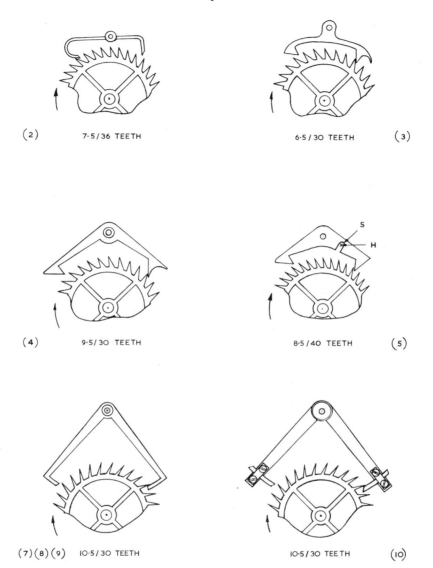

Fig. 59. Escapement Variants

drop on the entry pallet; however, the exit pallet drop remains almost unaffected. To change the exit pallet drop, we must change the spacing of the pallets, to bring them to the correct x-and-one-half tooth space distance.

The recoil escapement of Fig. 58a (recoil no. 1) is found in most American clocks, including the ogee, and other typical weight- and spring-driven clocks. It is designed well and economically; it is dependable and easy to adjust; and access to it is usually excellent.

To set up such an escapement, first remove any pits in the pallets with an oilstone or polishing stick. Remember not to grip a hard-steel verge too tightly in the vise, or it may break. If you have softened the verge for bending, file out any deep pits prior to polishing. The working edges of the pallets must be square and sharp, tapered only at the back.

Fit the verge in place, and check the play on the verge pin. If the holes in the saddle are too big, the verge will be loose; this must be corrected. If the saddle holes are spurted, they can often be hammered down on an anvil—provided that a suitably thick piece of steel is first pushed between the lugs of the saddle. These two holes, thus reduced, must then be broached again to ensure roundness and a good fit on the pin.

If no spurting is present, the best remedy for oversize holes is to remove the pin and replace it with a larger one, broaching the holes to suit. Yet another solution is to make a new saddle, with smaller holes which can then be broached to size.

When the verge is a good fit on its pin, raise it by means of its adjustable arm almost to the point of preventing the escape gear from running. At this stage, the exit-pallet drop will almost always be too great. This happens because the polishing of the pallets and sharpening of their ends has moved them slightly apart. To remedy this, remove the verge, and then, making sure that it is soft, use two pairs of long-nosed pliers to gently bend it, over the QR section (Fig. 58a), so that the pallets are brought a few thousandths of an inch closer together. Do not use much force; only a small shift is required, and the necessary change in shape will hardly be discernible.

Now replace the verge, and compare the drops again. Keep the verge as close to the escape gear as possible and adjust the pallet distance again if necessary. Repeat these adjustments until the drops are equal and as small as is consistent with reliable operation.

Among the faults that may frustrate close adjustment of the escapement are bent or incorrectly sloped teeth, play in the escape-gear pivot, an eccentric escape gear, or a loose saddle. Although bent teeth should have been straightened in previous steps, it sometimes takes a running escapement to reveal them. Run the escape gear manually, using a loupe to observe its action. A small hook at the tip of a tooth, a tooth which leans forward or backwards, a tooth which is a little too short—all cause trouble. Bent teeth can simply be pulled straight with pliers, using a slight bending action; a short tooth, however, must be replaced (see **Tooth replacement: Gears**).

Any eccentricity in the escape gear is most unwelcome. The gear may run normally over some two-thirds of a turn, and then refuse to go further. The remedy here is to "top" the offending teeth and then file them sharp again. Topping—meaning simply shortening—can be done between centers in a lathe; however, by the time this problem is perceived, the clock has usually been fully assembled. It is therefore simplest to let the movement itself provide the necessary rotation.

Take a fine parallel file and lay its thin edge on the clock plate, with the flat cutting surface resting lightly but squarely against the escape-gear teeth. Pressing one end of the file firmly against the clock plate, move the other lever-fashion to apply light but firm pressure on the teeth, while spinning the gear by hand via the time-train gears. If the movement is spring driven, simply wind it up and allow it to run down, moving the file in small increments to reduce the high teeth. It is essential to keep the cutting surface of the file at 90° to the plane of the gear, or the tooth tips will not be square. The file must be held very steadily, and must cut only the high teeth until they are level with the others. In an internal escapement, the file can be rested against a convenient arbor, and so automatically keeps the tooth tips square.

When all the teeth are level to the eye as the gear spins, file the leading edges—those which contact the pallets—until all the teeth are uniform. Carefully remove all burrs with a fine needle or escapement file; an illuminated magnifier is very useful for this. Smooth the tips of the teeth lightly with very fine emery cloth. Readjust the verge, if necessary.

A loose saddle should be re-riveted on a narrow anvil, using a flat punch.

When the verge is satisfactory, harden the pallets (as described under **Hardening**) and repolish them. If the crutch is loose in the verge, clean the steel around its entry point and solder it tight; it is not easy to rivet it again. Naturally, all soft soldering must be done after hardening is completed.

Now put the movement in its case and make sure it is level. Set the clock in beat by bending the crutch. This is done by grasping the crutch at L (Fig. 60a), and moving the bottom of the crutch to the left or right. When the clock is vertical in the forward-backwards sense, the pendulum rod must hang in the loop of the crutch, as in Fig. 60a. The configurations shown in Fig. 60b and c are both incorrect, and cause a bind capable of stopping the clock.

The purpose of the oval loop is to allow freedom in the forward-backwards sense; the loop must therefore be loose about the pendulum rod but as close to it as possible in order to prevent lost movement. A tiny drop of oil is not out of place on the loop, and makes for quieter operation.

On completing an overhaul, a tiny drop of oil on each pallet helps the escapement run itself in. I prefer to use synthetic precision watch-and-clock oil. Aircraft instrument oil, another synthetic, also works well. Being on exposed surfaces, this oil will dry up within a matter of months, and the escapement normally will continue to run without oil. If the clock stops within a few weeks or months, however, it is usually due to a dirty escapement. The small black deposits are easily removed from the pallets and escape-gear teeth with a swab soaked in white gas.

The formation of this black deposit is evidently activated by the cleaning solution (see **Cleaning process**); it also can occur on other gears. To avoid this occasional problem, pro-

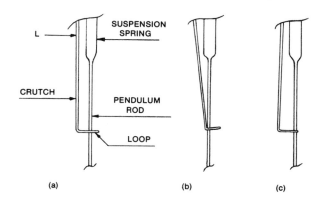

Fig. 60. Crutch Settings

longed rinsing during the cleaning process is indicated; an overnight soak for the escape gear is an excellent practice. Another approach is to clean the gear by hand with a toothbrush and cleaning powder, without immersing it in the solution at all.

Cleanliness is important in escapements, but having hard, mirrorlike pallets and smooth escape-gear tooth tips also plays a major role in the continuing operation of an escapement after the oil has dried up.

Recoil no. 2 (Fig. 59) also uses strip pallet or verge, but it is riveted into a slot in an arbor and so lies between the plates of the clock rather than externally. Such verges are often unhardened; however, this cannot be taken for granted. They should be adjusted in the same way as no. 1.

Recoil no. 3 (Fig. 59) uses a solid anchor, rather than a strip-type verge, but it is light enough to be adjusted after softening by being squeezed gently in a vise or opened slightly with pliers. It is not too common, being found in clocks such as the banjo. Usually only the pallets have been hardened, but it is always best to check the anchor for hardness before attempting to bend it. (This applies to all solid anchors.)

Recoil no. 4 (Fig. 59), being from an English longcase, is much the same as the original "anchor escapement" attributed to Robert Hooke or William Clement (ca. 1670). Although soon followed by other anchor escapements—such as all those listed earlier—the term *anchor* still means *recoil* to some people.

This escapement uses a somewhat massive anchor. It is not always easy to adjust this type sufficiently to compensate for the relatively deep pits sometimes to be found in its pallets. If the pallets are brought in sufficiently after the pits have been dealt with, they are sometimes no longer at optimum angles. Although purists evidently frown on the practice of soling the pallets, I consider that when properly done it makes a sound, repeatable and workmanlike repair, but should not be used when shallow pits can simply be polished out.

Check the anchor for hardness. If the arms are too hard to bend, remove it from its collet (it will often be soft soldered in place), and soften it throughout, as described under **Hardening.** If only the pallets are hard, simply soften them without taking the anchor off the arbor; heavy pliers make a good heat sink for this job.

To sole the pallets, first file both pallets true, so that the deep pits practically disappear. Flatness and squareness are more important here than reducing the pits to zero. Then clean up the sides of the pallets. Now, using heavy pliers, break off two small pieces from a piece of old 0.018″-thick clock spring, of a size to just cover the pallets, slightly overlapping them on all sides. Trim them on the grindstone, leaving them slightly oversize, and thoroughly brighten both sides with a coarse polishing stick. Both pieces must be flat; with small pieces taken from the outside coil of a broken mainspring this is not a problem, and slight adjustments are feasible.

Now, flux and tin both pallets and the worst sides of both soles. Solder them carefully in place, and then slowly trim all available sides on the grindstone; do not press hard or heat may melt the solder. Make sure to shape the undercut rear edges of the pallet tips correctly with an old round file. Hone both pallets flat with a kerosene-soaked oilstone, and then bring them to a fine finish with polishing sticks.

Fit the pallet arbor in place; check that the anchor is square with the escape gear and is overlapping it equally on both sides, taking end play into account. Slide the anchor along the arbor or straighten it, as necessary.

When well done, the above procedure will ensure pallets that are quite close to their optimum positions. However, small adjustments are still usually required.

There is no provision on English longcase clocks for changing the drop of the entry pallet, for the pendulum cock is pinned, as well as screwed, to the back plate. To obtain some movement, it is necessary to knock out these two pins and elongate the clearance holes in the pendulum cock. As the anchor generally needs to be lowered, the holes have to be filed out upwards somewhat.

This escapement is set up in the same way as no. 1. The pallets can be closed a fraction in the vise, and opened by gently squeezing either of the slightly curved pallet arms in a pair of long-nosed pliers; use the pliers close to the collet.

There is no real need to reinstall the two locating pins in the pendulum cock; the screws are entirely adequate. A further point must be noted: When the pendulum is set in place, its weight will cause the cock to sag a little, sometimes enough to cause the escapement to jam. If you wish to fit the pins again, it is best not to drill new holes for them until the final position for the pendulum cock is found.

Recoil no. 5 (Fig. 59), has a very solid anchor which is much too heavily made to be bent, even slightly, and rather too small to be easily soled. It is found on mantel clocks, including those with Westminster chimes. Presumably, it was the designer's intention that such anchors simply be replaced when worn, but since this is no longer feasible, other means must be found when deep pits cannot be polished out without spacing the pallets too far apart.

One approach is as follows: Remove the anchor from the arbor. Since it is usually hardened throughout, soften it thoroughly. File and polish the pallets, removing only the minimum of metal, and drill a hole in either arm, centered at H (Fig. 59, no. 5). Saw down to it to give a round-bottomed slot as at S. The anchor is now set up in the same way as in no. 4, except that a screwdriver is inserted in the slot when moving the pallets apart during adjustments. (The purpose of the hole is to avoid the damaging localized stresses present at the end of a sharp cut.)

After setting up the escapement, remove the anchor and harden both pallets, before completing the repair.

The deadbeat escapement (no. 6) in Fig. 58b is typical of the other deadbeat variants, nos. 7, 8, 9, and 10.

The deadbeat escapement came into use about 1730; it is credited to George Graham. The intent is to limit the pendulum to a small and constant arc, thus minimizing circular error (explained under **Pendulum**).

Note that the teeth of a recoil escape gear slope backwards, away from the direction of rotation, and that this slope is reversed in the deadbeat, the teeth of which also have at their roots a clearance space not always present in the recoil.

Each tooth of the deadbeat escape gear, instead of landing on an active pallet face and bouncing back a little, as in the recoil, lands on a curved surface which, although moving, does not influence the tooth either forward or backwards. It is this "dead" part of each pallet that gives the escapement its name. To allow the anchor to move without influencing the escape gear, the curved surfaces of the pallets have a radius equal to the length of the tangent from the verge pivot to the O.D. of the escape gear. Fig. 58b shows the two curved surfaces at T and U; the inset gives set-up data. In general, the deadbeat is set up along the same lines as the recoil; however, equalizing and depthing are more critical.

When the tip of one of the forwards-sloping teeth lands at point V (Fig. 58b insert) on the dead surface T of the exit pallet, the pallet has almost stopped rising. The return swing of the pendulum then causes the pallet to fall, and the tooth, which has not moved since contact was made, now slips over the edge of the pallet and slides across the impulse face X (inset), thus pushing the exit pallet downwards. Another tooth (spaced x-and-a-half tooth spaces behind) then lands on the dead surface U of the entry pallet, and a similar hold-and-release sequence occurs. The whole cycle then repeats itself.

Since the whole action is not as lively as a recoil, the V-points do not pit as severely, but the edge between the dead and impulse faces slowly loses its sharpness; some scoring also occurs on the impulse faces. The wear on the dead face can usually be polished out by a few strokes of an emery stick, but the impulse face requires the full treatment, for it must be flat, mirror-smooth, and at the correct angle.

This might seem like a tall order, but, as usual in such cases, a small jig is all that is required. Fig. 61 shows a small pair of steel chops which will hold the verge securely so that the impulse faces can be stoned and polished.

The two chops K and L, shown in Fig. 61a, are dimensionally alike; they only differ in that chop K has two recessed 8-32 N.C. clearance holes in it, whereas chop L has two through holes, tapped 8-32 N.C. The two holes must be drilled with K and L clamped together, so that the top surface M (Fig. 61b) is perfectly flat. The screw heads must be well inset, and neither end of the screws must project when K and L are screwed together; this enables the jig to be mounted in a vise, after the verge or pallet has been carefully set up in it. In most cases the direct use of a vise, without a jig, does not work well, due to misalignment, serrations, and the lack of a flat, smooth top; also, a vise is usually too large and clumsy.

Set the verge up in the chops, as shown in

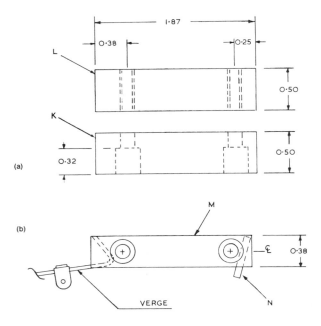

Fig. 61. Pallet Jig

Fig. 61b; a spacer N, equal in width to the verge, must be used to maintain parallelism. The exact angle at which the verge lies is determined only by the existing impulse face, for this angle must not be changed, unless it is obviously incorrect. The impulse face must project upwards from the top of the chops enough to enable the scoring and rounding to be polished out; usually about 0.004″ is enough, with bad cases requiring 0.006″ or more. It is fastest to use a kerosene-wetted oilstone initially, followed by a smooth polishing stick; both of these are laid flat, with the aligned top surfaces of the chops acting as a guide.

This jig can be used for other deadbeat pallets, their anchors being set up between the screws. Adjustable Vienna pallets must be removed from their anchor before being mounted in the jig. Excessive pressure from either jig or vise can crack a hard verge or anchor, so care is necessary in clamping. The jig does not take long to make, and gives excellent results. It is not advisable to attempt to resurface deadbeat pallets without some form of jig to hold them.

The deadbeat verge must be adjusted in the same way as the recoil (p. 156), being softened as required. However, the two drop adjustments interact rather more in the deadbeat, so it is a little harder to adjust. Basically, the two drops must be equalized, and at the same time the tip of each tooth must land at point V (Fig. 58b insert) on the curved dead surfaces of the pallets. The point V is identically placed on both pallets. The distance from it to the corner of the pallet equals one-third of the width W of impulse face X.

When the verge is correctly set up, harden it as before, and repolish both pallets to a fine mirror finish. The saddle and crutch are dealt with as in the recoil escapement.

The design of deadbeat escapements nos. 7, 8, and 9 (Fig. 59) is a very typical solid pallet version of type no. 6 (Fig. 58b). It is found mostly in clocks of the regulator type. The pallet arms are usually malleable, and so do not require softening—but this must always be checked. The pallets themselves are, or should be, hardened.

There is generally no need here to soften the pallets. When worn, they respond very well to the refacing treatment outlined for no. 6 above, and the same jig can be used. The relatively slender pallet arms of the anchor move fairly easily in the vise or pliers; but, since this setting is quite critical, several adjustments are often required before the exact spacing is found. The set-up criteria are the same as for no. 6.

The depth of the anchor can usually be adjusted by means of a small, slotted turntable, or equivalent device, which carries one end of the pallet arbor. As this adjustment is small, it does not usually misalign the rather narrow pallets, on which the escape gear is centered. This is always worth checking, as is the end play on both arbors. Because all deadbeat escapements are more critical to adjust than recoil types, it is unwise to try to hurry the process.

Deadbeat no. 10 is used in Vienna regulators and other superior-quality clocks. The same setting points are required as for nos. 6–10, but adjustment is somewhat easier, as the pallets themselves are movable in the anchor.

The small, curved, hard-steel pallets are cut from a circle, as mentioned earlier (see page 161), and they lie at the ends of the anchor arms in trunnion slots, in which they are clamped by means of a strap and two small screws. Loosening the screws allows each pallet to slide, and this greatly helps in adjusting the distance between them to obtain the correct exit-pallet drop. Here there is no need to bend the brass anchor, unless it happens to be warped.

If the pallets are worn, remove them and resurface them in the jig as previously described; do not soften them. In many cases, the pallets are sharp at both ends and so can be used twice. They cannot simply be changed end-for-end, however; they must also exchange positions, for otherwise the impulse surfaces would be reversed. With this reversible design only one pallet had to be made, which cleverly simplified production.

The anchor is generally raised and lowered by a slotted turntable. The crutch pin (at the bottom of the crutch) is movable by means of a screw. The clock can thus be set in beat without any bending of the crutch wire, thus minimizing the chances of damage to escape-gear teeth fine pivots. The crutch pin must sit firmly against the bottom of the crutch; it may be necessary to adjust the two lugs that hold the beat-setting screw in order to tighten it up. The thick base of the crutch pin must not push against the pendulum; if it does so, bend the top of the crutch gently towards the back plate. The brass plate on the pendulum rod can be squeezed slightly in the vise to narrow its slot a few thousandths of an inch, if the crutch pin is unduly loose in it; always make sure that the pin is still free to move, however.

The weight of the beat-setting mechanism, placed at the end of a relatively long crutch, seems an unduly heavy load to be moved by a weak impulse, but it is part of the basic plan to limit the pendulum arc in these excellent timekeepers.

The Brocot escapement, nos. 11 and 12 (Fig. 58), is usually a deadbeat variant, but if an escape gear with backwards-sloping teeth is used, it becomes a recoil escapement.

The pallets here are semicircular pins made of either stone (often carnelian) or steel. Some may have had additional flats ground on them (see the dotted line in the insert for Fig. 58c), thus making them very close to the Graham pattern. This sometimes has to be done to overcome the Brocot's tendency to stall. Aside from perhaps using more steel than stone in the pallets, the American version scarcely differs from the original nineteenth-century French Brocot; Fig. 58c serves to illustrate both.

When the pin pallets are worn, it is best to replace them rather than try to polish them. As the anchor is almost always of brass and not too heavily made, it is fairly easy to adjust the drop on the exit pallet. The Brocot escapement is often visible in the clock dial. This is a good place to show off the rather unusual teeth of the escape gear and the ornate bridge which forms the escape cock. Slightly changing the curves of this bridge permits the setting up of the entry-pallet drop. Both adjustments must be carefully done, as any marks will show up clearly.

There is little choice, with the usual unmodified Brocot, but to set it up as shown in the insert to Fig. 58c, where the tip of the tooth extends only very slightly past the halfway mark on the pallet. The problem here is that the usual, short, light pendulum cannot always unlock the pallets when the mainspring is fully wound. If, on the other hand, the tooth tip lands short of the halfway mark, an uncertain action will again result. Although most clocks will run, a few may have to be modified as shown in the insert. The more successful versions utilize relatively short pallet arms which embrace fewer teeth; the pendulum bob should be as heavy as possible.

After setting up, clean the escapement carefully. If oil is applied, it tends to leave the pallets and collect at their bases. This can largely be prevented by the use of undercutting, but the problem of evaporation still remains: Dried-up oil still stops these clocks in the same way as with other deadbeat variants; but since the escapement is readily accessible, it is easily cleaned up with white gas.

Fan or Fly: a rotating vane which uses the surrounding air to regulate the running speed of a clock's striking or chiming train. Some fans are combined with small centrifugal weights to reduce their size.

Fly: *American and German.* To make a fly for an American or German clock, you will need a simple forming jig to press out the half-round, central channel in the fly (see Figs. 62 and 63).

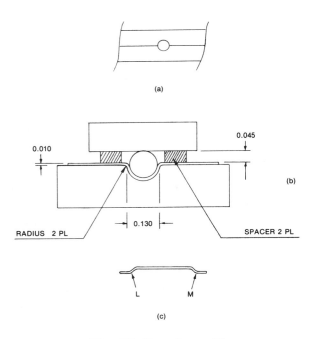

Fig. 62. Forming a Fly

Such a jig is easily made by clamping two pieces of steel together and drilling a hole down their mating surfaces (Fig. 62a). Typical dimensions for the steel blocks are 1.5″ × 1.5″ × 0.375″.

The drilling must be done on a drill press; it is best done from both sides of the blocks. Scribe a line around the blocks, using a square to denote the drilling points. Separate the blocks after drilling, and smooth out each channel with a rattail file. Then round the channel's edges slightly to allow a small radius when forming the groove in the fly (see Fig. 62b—radius, 2 pl.).

From 0.010″-thick brass shim stock, cut a blank that is slightly oversize relative to the intended final dimensions, and align it centrally over the channel in one of the blocks. This blank will become the fly. Lay a suitable small steel rod on the brass blank above the channel, and set two small spacers of appropriate thickness on either side of it; two-sided masking tape will hold these in position. Lay a flat block of steel on the rod, and then press the whole assembly firmly together in the vise. A groove will form in the blank, which can then be trimmed to the required dimensions. A final press without the masking tape will improve definition.

Fig. 62b shows the general procedure and gives typical dimensions. Now drill two 0.024″ holes in the fly, to line up with the groove in the fly arbor, and bend a short piece of 0.020″ spring steel wire into them so that it runs in the arbor groove and provides tension. A suitable wire shape is shown in Fig. 62c; the two bends L and M can be made and sharpened up after insertion in the fly. Slide the arbor along the channel in the fly until the wire clicks into the groove in the arbor.

Fly: *English longcase.* The original flies were evidently made by the time-consuming and wasteful method of filing them out of solid brass some 0.220″ thick. Because rolled brass did not appear until about 1825, there was possibly little choice in the matter. But since

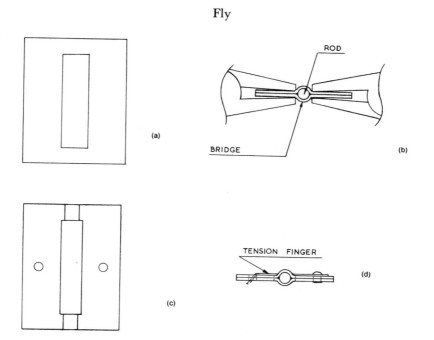

Fig. 63. English Longcase Fly

flat metal sheet is available to us today, it seems sensible to use it.

Before cutting any shimstock, consider the O.D. of the bushings. An old fly arbor usually tapers, and typically has a maximum diameter of about 0.110″. A suitable bushing, 0.180″ long, with a 0.035″ wall thickness, will therefore have an O.D. of 2 × 0.035″ + 0.110″ = 0.180″. The width of the slot in the flat blank of Fig. 63a will therefore be 0.180″ × π ÷ 2 + 0.06″ = 0.35″ ($^{11}/_{32}$″); this figure includes 0.03″ each side as a bending allowance. The length of the fly (L) will be determined by the length available on the fly arbor, hence the length of the slot in the blanks will be L − 2 × 0.180″ or L − 0.36″. The blades of the fly will be as wide as nearby arbors allow.

Bearing the above data in mind, cut out two slightly oversize blanks from 0.010″ brass shimstock to the shape shown in Fig. 63a. These blanks will become the two halves of the fly. The slots can be cut out with a hand notching tool or chain drilled and filed. Over-sizing allows for trimming—note that the slot will need to be slightly *shorter* than its final dimension.

No. 16 or no. 15 drill shanks or stubs will serve for the 0.180″-diameter rod; use 0.080″-thick spacers and a channel 0.100″ deep for pressing out the grooves (bridges) in the ends of the blanks, as shown in Fig. 62b. Follow the procedure given in the previous section, taking care to align each blank with the channel in the block.

While the jig method gives excellent results, it is also feasible to use instead two large pairs of pliers (Fig. 63b). The two bridges are quite narrow and so can be bent, one at a time, around the drill shank. Although it requires care to prevent slippage and keep the bridges in line, this method saves making a jig.

Make sure that the two halves of the fly are flat, and can contact each other, closely and entirely, with the rod in place. Clamp them in this position. Drill two 0-80 clearance holes (no. 52 drill) in the vanes as shown in Fig. 63c.

To do this, position the arbor correctly in the bridges, and drill the two holes opposite the groove in the arbor. Note this position; the arbor will not fit if inserted from the other end, because in almost every instance the groove will no longer be opposite the holes. Unclamp the two halves of the fly and thoroughly clean, flux, and tin their mating surfaces and the insides of the bridges.

Turn up the two 0.180″-diameter brass bushings, making them as long as the width of the bridges (nominally 0.180″). Make the I.D. of one bushing about 0.005″ under the minimum arbor diameter and the other bushing 0.005″ under the maximum arbor diameter. Both holes will be broached out later. Let us call the bushing with the larger hole in it bushing "E."

Screw the two halves of the fly together. Position the arbor correctly in the bridges, as before, to check that the groove aligns with the holes. Now withdraw the arbor and insert the bushings accordingly; bushing E will be at the entry side of the fly, close to the arbor pinion.

Now solder the two halves together, including the bushings, using the corner of a block of wood and a 100-watt soldering iron. Heat each corner in turn, and press the two halves together with a wooden dowel until the solder sets. Take out the two screws and clean up any flashings of solder and brass. As necessary, correct the shape of the slot and the outline. Broach out the bushings, from the entry end, to take the arbor.

Then enlarge the two screw holes, as required, to take a slim tension finger about 0.015″ thick (shown in Fig. 63d). The free end of the finger should be cranked where it passes through one hole; the fixed end is then riveted into the other hole. A fly made by this method will not look too unlike the original.

Friction collet: a collet mounted on an arbor, such as the winding arbor of an English longcase, to keep the drum tight. (See page 75.)

Fig. 64. Gear Expander

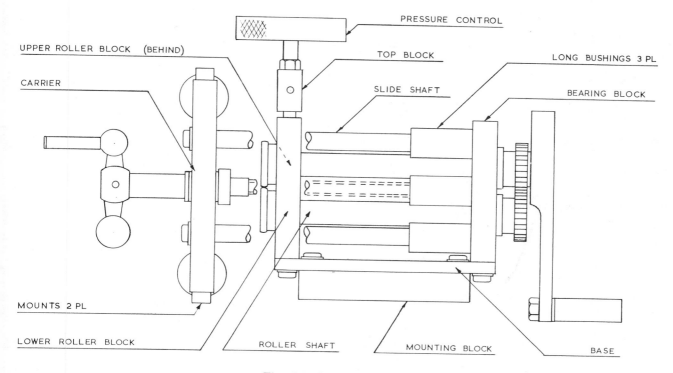

Fig. 65. Gear Expander

Friction spring: a spring washer, usually mounted on an arbor, to apply a little pressure to a count gear or winding drum. (See page 26.)

Gathering pallet: A pallet or pin which counts—i.e., gathers—the teeth of the rack in a rack-and-snail striking mechanism (see Fig. 26).

Gear: a toothed wheel, as in a clock.

Gear expander: Owing to wear or, occasionally, poor initial layout, the gears of a clock do not always mesh correctly. In some cases, to make the clock run it may be necessary to move a pivot hole or two; this can be a slow and troublesome business when the gear meshes with two others. However, it may be the only option. As the pattern of wear usually results in too light a mesh, and only one gear is generally involved, a little gear expansion is often the easiest solution.

Although it takes many hours, constructing a gear expander is an interesting challenge—and a useful machine results from the effort. The expander of Fig. 64 will handle arbor-mounted flat gears 1″–3″ in diameter and up to 0.060″ in thickness. This covers a large percentage of common clock gears. The gear arbor is held vertically between two pre-set brass mounts, which can slide in a screw-driven carrier to adjust the position of the gear. This arrangement allows the gear rim to turn between two pressure-adjustable steel rollers. The roll path can be changed during normal rolling to cover the spokes or any part of the rim.

Fig. 64 shows the completed machine, with a gear in position for rolling. Fig. 65 is an assembly drawing drawn to scale. The chief components are detailed in Figs. 65 to 69.

Gear Expander

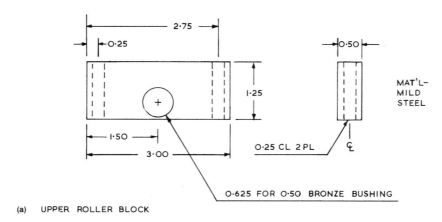

(a) UPPER ROLLER BLOCK

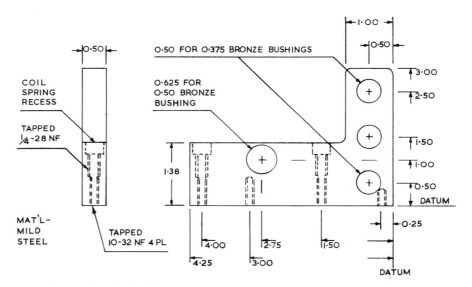

(b) LOWER ROLLER BLOCK

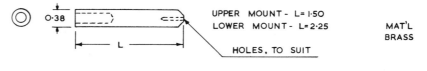

(c) MOUNTS

Fig. 66. Upper and Lower Roller Blocks and Mounts

Gear Expander

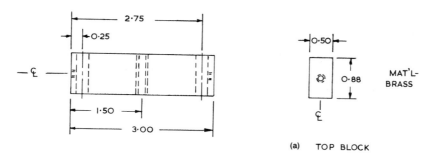

(a) TOP BLOCK

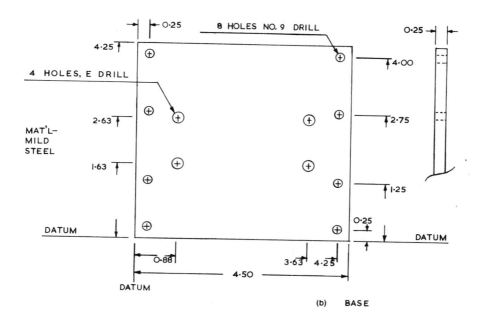

(b) BASE

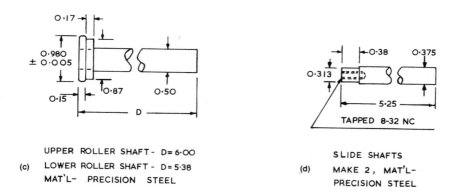

(c) UPPER ROLLER SHAFT – D = 6·00
LOWER ROLLER SHAFT – D = 5·38
MAT'L – PRECISION STEEL

(d) SLIDE SHAFTS
MAKE 2, MAT'L – PRECISION STEEL

Fig. 67. Top Block, Base and Roller Shafts

All flat parts here are first cut from stock on a band saw. It is then preferable to true edges and drill all critical holes in a milling machine, but when no mill is available the edges can be filed and trued by hand and the holes drilled on a drill press. This may raise a few machinists' eyebrows, but is nevertheless the method I used in making the expander shown. As there are eight critical holes, which involve two shafts and a lead screw, great care is required in their placement; unless the reader is experienced and skilled in handwork, it may be just as well to have the holes machined by a competent machinist.

The mounts (Fig. 66c) are double ended: One end carries a fine hole to take the small arbors of 1″ gears; the other has a larger hole for the pivots of winding arbors, which may carry gears almost 3″ across. Custom mounts can be quickly turned and drilled on the lathe.

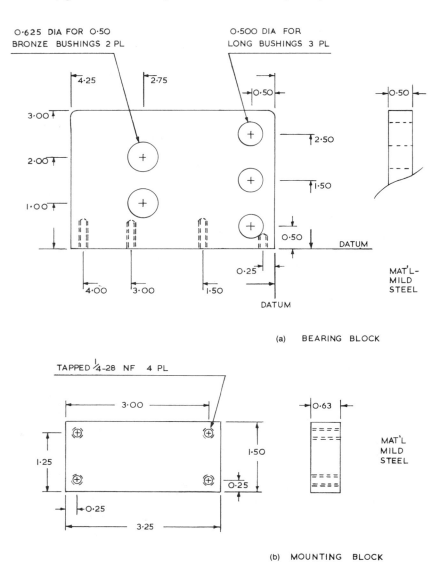

Fig. 68. Bearing Block and Mounting Block

Two knurl-headed setscrews hold the mounts in position.

The roller shafts and slide shafts should be of reasonably hard, well-finished steel such as stainless or other precision shafting. The roller shafts are turned down at one end to a drive fit for the two rollers, which must be of harder steel than the shafts—e.g., Keewatin or a similar tool steel; tempered drill rod is also acceptable. The two roller shafts run in oil-impregnated bronze bushings, and the normal play in those of the bearing block easily allows the rollers their 0.060″ separation. This arrangement does not disturb the two 1″ driving gears.

The long bushings of Figs. 69b and c are made from ⅝″ stock brass; section TU is a press fit into the bearing block of Fig. 68a. In Fig. 69b section RS is a bearing fit for the ⅜″-diameter steel slide shafts, and section SU is a

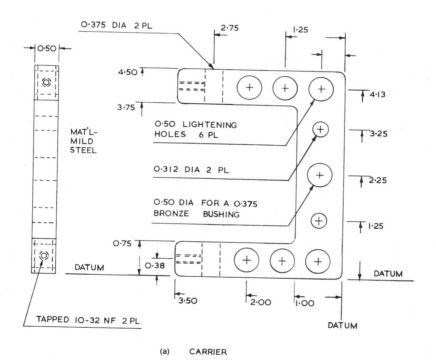

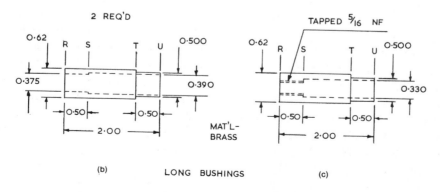

Fig. 69. Carrier and Long Bushings

clearance diameter—say, 0.390″. In Fig. 69c section RS is tapped 5⁄16-24 NF, or a similar thread, for the carrier lead screw; section SU is again a clearance diameter. The use of long bushings avoids unsightly overhangs, and also protects the lead screw and the slide rods from accidental damage.

The lower roller block of Fig. 66b carries two ¼″-diameter, steel rods, 3½″-long, each end of which is threaded ¼-28 NF for ½″. The two rods screw into holes in the block. Recesses are provided for two small coil springs which keep the rollers apart by pressing the upper roller block upwards against the fixed top block. The upper roller block slides on the rods; the top block is secured to the rods by locknuts and 8-32 setscrews. The top block is of brass, to better accommodate the ¼-28 NF steel pressure screw that passes through it, and is turned by a knurled knob or suitable spare gear.

The mounting block of Fig. 68b is not of critical dimensions, being simply for clamping the expander in a vise.

As end thrust is not great, either flanged bushings or flat brass washers may be used throughout. A 3″ crank is used for turning the rollers, while a 1″ counterbalanced handle is suitable for the carrier lead screw.

When complete, the expander should be rustproofed and painted; the inside of the lightening holes in the carrier can be painted a contrasting color. The steel shafts and long bushings should be polished and the latter lacquered, thus making a handsome machine. It should be oiled periodically.

To use the expander, clamp it in the vise, with the roller crank to the right; measure the gear to be expanded, and set it up in the mounts, raising or lowering them until the rim of the gear fits between the rollers. Check that the gear is level and resting comfortably on the lower roller with its teeth beyond the rollers in the clearance provided. *Do not roll the teeth themselves,* as this will merely flatten them and cause problems.

Bring the top roller down onto the gear with the pressure screw, and check the rotation with the roller crank. If the roll path is correct, apply a little more pressure as you turn the crank, and then move the carrier in and out so as to cover all of the gear rim, cranking meanwhile continuously.

Do not apply much pressure to the rollers; it is quite easy to apply too much, and so expand the gear beyond help. As the gear expands keep measuring the diameter with a micrometer or vernier caliper. Roll the rim evenly to preserve concentricity, and give each spoke a turn or two, as required to keep the gear flat. Any eccentricity can be reduced by lightly hammering the appropriate spokes afterwards, but this problem does not usually arise unless excessive pressure is used.

The whole concept is to expand the gear, slowly and evenly, perhaps 0.020″–0.025″, to improve the mesh with other gears; more than this is seldom required. The machine is easily capable of rapidly expanding a gear by several times this amount, so it must be used with care. An adjustable finger may be added, to lie close to rest the tips of the teeth and so provide a check on the diameter during rolling.

Glass cutter: *Circles.* Because circles of glass are frequently required for clocks, it is desirable to be able to cut them out. The common six-wheel glass cutter can easily be adapted to cut circles, as shown in Fig. 70. First take a 9.5″ length of steel I-beam curtain track, and drill eight or nine holes (no. 18 size), 1″ apart, in the center of the web. From 5″ of coathanger wire form a U shape, ½″ wide, that will fit snugly in the curtain track. Some ⅝″ up from the bottom of the U, bend both legs 90°. Now, with a no. 18 drill, make a hole in the glass cutter as shown in Fig. 70, and screw the U-wire to it, using a washer and lock nut.

Make two mild-steel chops ⅛″ × ½″ × ¾″, and drill a central hole in one of them with a no. 18 drill. Drill the other with a no. 29 drill, and tap it out 8-32 NC. Clamp the limbs of the

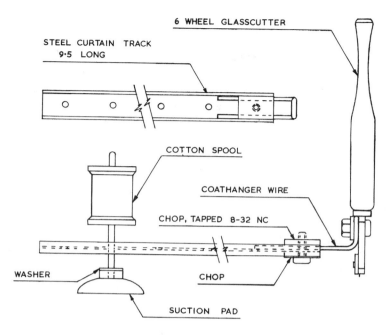

Fig. 70. Glass Cutter: Circles

U-wire between the chops as shown in Fig. 70; this makes possible a fine adjustment of the cutting radius.

Obtain a small, circular rubber suction pad, about 1.5″ in diameter—a map fence, for a car dash, is a good source—and embed the head of a smooth-shafted 8-32 screw in it so the shaft points upwards. Slide a washer of less than ½″ O.D. down the screw shaft. As a hold-down for the suction pad and curtain track (Fig. 70), use a wooden cotton spool or a 1″ length of 1″-diameter birch dowel with a hole (no. 16 drill) in it.

To use the cutter, set up the suction pad in the hole in the curtain track closest to the required radius, and set the cutter up on a thick pad of newspapers. Make two wheel marks 180° apart and measure their distance (the diameter of the circle). Loosen the chop screw to adjust the U-wire until the correct diameter is obtained; and then tighten the screw again. Put your glass on a few thicknesses of newspaper laid on a clean, flat bench, and set the cutter in place. Press the suction cup down via the hold-down, and keep it there; now cut the circle, using firm pressure on the cutter.

Tap round the cut from underneath with a screwdriver handle, and see that the cut runs as intended; a few tangential cuts with a normal glasscutter will help the breakout.

Gong rod: Fig. 71 shows the usual form of a gong rod. It is musically sound but mechanically unsound, which accounts for the fairly high breakage rate. Gong rods always break off at their weak point—point D in Fig. 71—generally due to accidental bending during transporting or setting up the clock.

In a bim-bam strike it is often feasible to machine the broken rod's other end and reverse the rod, leaving the pointed end free. As this shortens the rod and so raises its tone, it may be necessary to exchange the positions of the two rods and retune them by shortening the other one a little. Rods may be wedged or screwed into the gong base.

In a chiming clock, the rods form a musical scale, so a broken one must be replaced by one

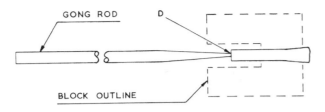

Fig. 71. Gong Rod

of the same tone. This is done by installing an extra long rod and cutting it down until the desired note is achieved.

To prevent such breakage, it is essential to secure the free end of all gong rods to the case before disturbing the clock in any way. Many clocks have clamps for this purpose, and they should always be used.

Grand sonnerie: a European system whereby the quarter-hours are struck on one gong, followed immediately by the striking of the last hour on a second gong.

Hand collet: A collet is more substantial than a washer, and the type of clock-hand retainer used on English and European clocks is rather too solid to be called a washer. A few American clocks also use such collets. Where they are used, they form part of the hand-clutch system, being under constant pressure from the hand-clutch washer, whereas the American-style hand washer simply holds the minute hand on the arbor.

It is sometimes necessary to "dish" such a collet more deeply—that is, increase its concavity or convexity in order to increase the pressure on the minute hand. To do this, drill a shallow hole in a piece of hardwood, such as birch, making the hole a little smaller in diameter than the collet. Lay the collet, convex side down, on the hole, and set a steel ball from a ball race—of about 5/16" diameter—in the hollow of the collet. Now rest the end of a short brass rod, some 1/2" in diameter, on top of the ball, and tap it smartly with a small hammer.

Instead of a wooden block, an anvil—with a suitable hole in it—may be used, but its sharp edges may mar the collet somewhat. The brass rod is used because a steel ball from a ball race is very hard indeed and would damage the face of the hammer if struck directly. This method approximates the use of a ball punch and dapping block; it is given because these items are not commonly at hand.

Hand washer: As indicated in the previous section, the term "hand washer" is associated more with American clocks than European ones. It denotes a concave-convex washer, pressed out from sheet brass about 0.015" thick. The central hole is round, square, or a slot, as required—the square being the most common.

Unlike the hand collet, the hand washer does not form part of the hand-clutch mechanism; in conjunction with a pin, it simply holds the minute hand in place. Since the hand sometimes becomes loose due to handling, the washer must be adjusted to take up the slack. The dishing can be increased quite easily by applying a pair of long-nosed pliers to four points, 90° apart, around the edge of the washer. When the washer has a square hole, the sides of the square can simply be bent further down, one at a time. Gently squeezing the washer with pliers close to the pliers joint will flatten it again.

Hands: *Repair.* The two chief types of clock hand are the pressed-steel variety, such as spade hands, and the more elaborate, hand-filed type which is found on old English longcase clocks. Both types can break in various places. It is wise to keep all broken pieces for repair purposes. Steel hands may be silver soldered, but brass hands must only be soft soldered, to preserve their hardness.

Fig. 72a shows a spade hand, broken typically at its base. If the hand can stand shortening by 0.10" or so, file a small slot in the disc, as shown, in which the broken shaft will fit firmly; then clean, flux, and silver solder it back together. Soft solder is inadequate for this repair.

Fig. 72b shows a typical American hand with an enlarged and ragged mounting hole. This can be repaired by soft soldering the intact disc from another broken hand underneath it, taking care to line up the square accurately, as shown in 72a.

Fig. 72c is a side view, showing where a new tip (from another broken hand) has been soft soldered onto a broken shaft; the visible surface is flat and the repair cannot be seen. The joint is carefully filled in with solder, and the whole hand repainted on both sides.

Fig. 72d shows how to repair the pressed-steel hand of a German box clock with a loop of 0.040" steel wire. The loop is formed to a snug fit around the collet, filed slightly flat on both sides, and soft soldered in place, with the legs securing the broken shaft. To synchronize the strike on the hour, the loop solder—and only the loop solder—is melted, to enable to collet to be turned. Gripping the legs of the loop with pliers will provide an adequate heat sink. A small driver in the square hole will turn the collet. This repair, although not visible, makes the hand far stronger than it was originally.

Fig. 72e shows how a wire fork, or a Y-shaped piece cut from 0.040" sheet steel, can be soft soldered underneath a brass longcase hand to make a strong repair. If 0.040" steel wire is used, after forming, it should be filed somewhat flat on both sides as in Fig. 72d; better surface-to-surface contact is then achieved, and a tidier job results.

Note that all of the above repairs depend upon scrupulous cleanliness prior to soldering; wire must be brightened with emery cloth before bending. (See under **Soldering**.)

Hardening: *Brass.* Cast brass—that is, brass poured when molten and allowed to set—when cooled, is in the annealed (soft) state. If this brass is then hammered, rolled, or bent

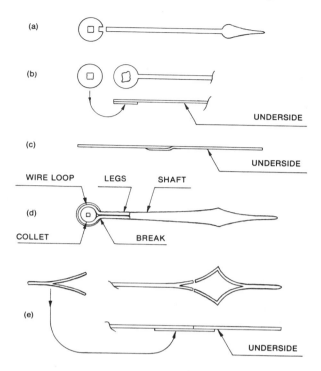

Fig. 72. Hand Repairs

into other shapes, it becomes harder; however, as this work-hardening, as it is known, always involves distortion, it obviously cannot be applied to finished objects. This is why it is not feasible to silver-solder gears made from rolled brass, for when work-hardened brass is heated above some 800°F., it simply reverts to its annealed state. This could easily cause the gear teeth to bend and fail, with disastrous results to the clock.

Hardening: *Steel.* Like most metals, mild steel will work-harden. It can also be case-hardened, by being heated in the presence of carbon; this forms an overall hard skin a few thousandths of an inch thick.

Carbon steel contains about 1 percent carbon. High speed and other special-purpose steels contain carbon plus various proportions of chromium, cobalt, manganese, molybdenum, nickel, tungsten, vanadium, etc. Carbon-containing tool steels are of enormous significance in engineering—and in clockmaking.

If mild steel is heated red hot and then quenched in oil or water, its hardness is virtually unchanged; if tool steel is so treated, it becomes very hard indeed—dead hard, as it is called. In most cases it also becomes brittle. Tool steel may vary in hardness, according to the heat treatment it has undergone. In the soft, or annealed, state, it is somewhat harder and tougher than mild steel, but is still malleable: It can still be bent, drilled and machined, and so on, with varying degrees of difficulty, according to its content.

Because dead-hard tool steel is generally too brittle to be useful, it is necessary to increase its toughness, even at the cost of decreased hardness. The process of doing this is known as tempering. Clock pallets are among the few items that can be used when dead hard, since the force applied to them is so weak their brittleness does not matter. After adjustment in the annealed condition, the ends of the verge (the pallets) are heated red hot and quenched in oil, leaving them dead hard—not tempered.

Tempering involves several degrees of hardness that lie between the extremes of the annealed and dead hard states; however, attaining the dead-hard state is almost always a prerequisite to tempering. The accurate tempering of many dead-hard items in mass production involves immersing them in mediums of known temperature, until they become equally hot, and then quenching them. Molten tin-lead alloys are very useful mediums in this regard, for example.

Fortunately, bright steel changes color over the tempering range, and this can be utilized in tempering workshop tools such as chisels, punches, and screwdrivers. Tools should not all be treated alike, but—as space limitations here do not allow the full, detailed treatment given in engineering handbooks—only the basic method can be given here. Table 5 shows the colors that appear as bright steel is heated—whether it is hard or not. It must be clearly understood, however, that tempering cannot occur unless the steel has first been made dead hard.

Finer gradations of color are sometimes given, but variance in lighting and people's eyesight renders exact description impracticable. Experience is essential in interpreting the colors.

The **Hollow punches** described later under that heading are typical of small tools that require tempering. Since only the hollow end needs treating, hold the solid end in vise grips, heat the hollow end red hot, and then quench it in oil or water. Quenching in water is faster, but induces greater stresses.

Now clean off the oil, and polish the steel bright again with emery cloth. The punch is now dead hard, and could well chip if used or dropped onto a hard surface.

To temper it, hold it as shown in Fig. 73 and apply moderate heat to the point indicated. Although it is required to temper only the tip of the punch, if heat were applied directly thereto, the colors would pass too swiftly to permit quenching. Therefore it is applied further up the shaft. As heating progresses, the steel at point P will slowly turn from its bright steel color to pale straw which will deepen to yellow-brown, as heating continues.

Now move the flame slowly back and forth as the arrow in the drawing indicates, stroking the color down to the tip with a rotary motion and turning the punch meanwhile, for even heating. When section PQ becomes an even

Table 5. Temperature Colors of Steel

450° F.	pale straw
470° F.	dark straw
500° F.	yellow-brown
530° F.	light purple
550° F.	dark blue
640° F.	light blue
800° F.	barely red
1600° F.	cherry red
2500° F.	white hot

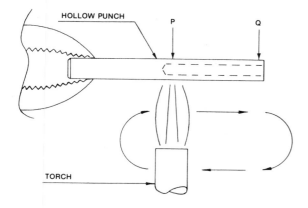

Fig. 73. Tempering

yellow-brown, quench the punch immediately in oil and hold it there until cool. The punch must enter the oil vertically, to prevent distortion. (The type of oil used is not critical; almost any general-use oil will serve.) Clean the punch, and check that you have caught the desired color. If you have applied too much heat and the punch is too dark or blue, simply repeat the whole process from the beginning, including hardening. The important thing is not to hurry: Remember to apply the heat slowly and to rotate the punch as it heats up.

Cutting tools should be left harder—at the straw stage—whereas screwdrivers, which must be tough rather than hard, should be heated further, to purple or blue. Note that when grinding tools, you should keep them cool by continually dipping them in water, so that no colors appear; otherwise retempering may be necessary.

History: There is insufficient space for in-depth horological history in this book, but it seems desirable to give some of the main turning points from the last six centuries of painstaking endeavor, even if only to enable you to reply to the perennial question, "How old is my clock?" Many illustrious names from the past are still spoken of today, and it inspires respect, to say the least, to hold in your hand a device either three centuries old or invented that long ago.

Although far from comprehensive, the following list will perhaps give some perspective to horological development.

First mechanical verge clocks	ca. 1300
Tower clock, Salisbury, England (restored 1931)	1386
Mainspring invented in Germany	ca. 1500
Galileo: swinging temple lamp, pendulum control conceived	ca. 1600
Pendulum clock devised by Huygens in Holland	1656
Verge-and-foliot gives way to pendulum control from	ca. 1657
Longcase prototype made by Fromanteel in England	1658
Recoil escapement and crutch invented by Hooke or Clement in England	1670
Rack striking developed by Barlow in England	1676
Famous horologists of this era include Arnold, Earnshaw, East, Graham, Harrison, Knibb, Tompion, Quare, and Windmills—all in England—and Berthoud, Breguet, and LeRoy in France	
Break-arch dial introduced	ca. 1715
Graham's deadbeat escapement	1730
Harrison's chronometers, etc.	1761
White dials appear in English longcases	1770
American wooden works	1790
Rolled-brass production starts	1825
American brass movements: mass production	1840
American brass movements exported to England by C. Jerome	1842
Westminster tower clock designed by Lord Grimthorpe and made by Dent in England	1850
Brocot escapement and pendulum suspension in France	ca. 1850

American mass production of watches	ca. 1850
Junghans sent to America to learn techniques, thus founding subsequent German mass production	1860
English trade declines markedly; very few longcase clocks after	1870
American mass production in full swing; firms such as Ansonia, Gilbert, Howard, New Haven, Seth Thomas, Waterbury, and Welch	
American production began to decline or change	1900

Hollow punches: These most useful tools can be quickly and easily turned and drilled on the lathe. Typical dimensions are given in Fig. 74; other sizes are equally useful. After facing and chamfering one end, center and drill the other; a depth of 1½″ is suitable for most applications.

These punches are made of drill rod, and must be hardened and tempered, as described above under **Hardening.** They are used in replacing a gathering pallet, moving collets along arbors, repairing loose clicks, and so on.

Hour pipe: The pipe which rotates around the minute arbor and carries the hour hand.

Huygens system: The ingenious arrangement shown in Fig. 75a, whereby a single weight drives both the time and strike trains of a clock, is credited to Christiaan Huygens, the notable Dutch mathematician, physicist, and astronomer of the seventeenth century.

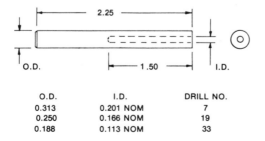

O.D.	I.D.	DRILL NO.
0.313	0.201 NOM	7
0.250	0.166 NOM	19
0.188	0.113 NOM	33

Fig. 74. Hollow Punches

The counterweight, which is usually a ring of lead, keeps the slack side of the endless loop taut enough to hold the loop on the pulleys, while the heavy weight—usually some 9 pounds—does the driving. One of the pulleys is on a ratchet to enable winding to occur. One advantage of this system is that maintaining power is built in, for power is not removed from the time train during winding. In England, clocks using this system far outnumbered the 8-day variety, because they were cheaper to make. The clocks have to be wound daily, running for about thirty hours between windings.

Either rope or chain was used for the loop. The pulleys have spikes, usually six in number, which prevent slippage and so provide drive. The joining of rope calls for a staggered splice, since the double thickness of a normal splice will not pass between the pulley sides. A disadvantage of rope is that it sheds fibers which, caught in the oil, are detrimental to the lubrication in the movement. Nowadays heat-joined plastic rope is sometimes used as a substitute, but the joint is always somewhat of a problem, whatever is used.

There is no such problem in joining a chain, of course. With a chain loop the pulleys have a slightly different configuration, and the spikes need not be as sharply pointed; however, they tend to wear to a point just the same. The rope loop runs quietly, but the chain gives out sundry bumping and grinding noises when it twists, as it inevitably does, while passing over the pulleys. Because chain systems are the most commonly used—rope-to-chain conversion kits tend to make them even more so—the chain system is the most important.

The wear on many old chains is in some cases almost enough to sever the links. As chain wear may cause these clocks to stop, sometimes permanently, it is important to analyze just what happens to them during normal running. Wear in the links (Fig. 75b) lengthens the chain appreciably, so that eventually there are only, for example, thirty-eight links per

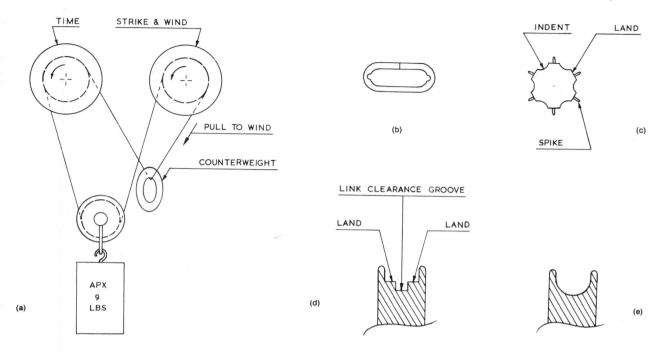

Fig. 75. Huygens System

foot instead of the original forty. Grooves which form in the ends of the links can also stiffen the chain so that it does not run as easily over the pulleys.

The pulleys' chain-bearing surfaces (*lands*), on both sides of the spikes, also wear down, as do the spikes, until the pulley becomes effectively smaller. A smaller pulley calls for more, rather than fewer, links per foot—yet, as mentioned above, the number of links per foot in the worn chain has decreased. The chain and pulleys, therefore, wear away in opposing directions and, in terms of the many years of a clock's lifespan, become incompatible relatively quickly.

More complications arise from the pulleys themselves. Generally, the separately driven no. 1 arbors of weight-driven clocks—as well as those of many spring-driven clocks—turn roughly the same number of times per day. In Huygens-system clocks this is unfortunately not the case, for the strike-side pulley, in addition to running the strike train, sustains all the stresses of daily winding, and therefore wears away from the chain even faster than the time-side pulley.

In view of this destructive wear pattern, it will no longer come as a surprise if the fitting of a new chain does not by itself cure the problem, for a chain that fits one pulley will not suit the other one. It is obvious, therefore, that the pulleys must be equalized before a new chain can be considered at all. If they are beyond redemption, there is no alternative to replacement. In that event, the most convenient solution is probably to use the above-mentioned conversion kit, which contains two pulleys and a matching chain. Any other old pulleys which may be available are very likely to be afflicted by the same disease, though perhaps to a lesser degree.

To preserve as much original equipment as possible, however, it is desirable to remachine both old pulleys to a common size, whenever feasible, and then to buy or make a new chain to suit them.

To do this, remove the no. 1 arbors and their pulleys from the movement. Punch out the pin that holds the strike-side pulley on its arbor, and slide the pulley off. Unpin the fixed, time-side pulley, and work it off the arbor. Examination will show that the sides of each pulley are pinned to the pulley's center section. Drill out and remove these pins, which will be lightly riveted, and then take off the two sides, marking each one for easy reassembly.

Check that the spikes are all present and equispaced; their positioning is critical. Occasionally they are unequally spaced, off center, or not set radially. Frequently they are grooved by long contact with chain links of slightly differing shapes. They are constantly involved in steel-to-steel confrontation with the chain; in fact, they lead a very hard life indeed. Frequently they are worn down so far that replacement is necessary.

The shortcomings of the Huygens arrangement have already been described; if it is to work even reasonably well, it is essential that the spikes of both pulleys be radial, equispaced, positioned on the center line, equal in length, and identical in shape. Although this may seem a tall order, short or displaced spikes stand out quite plainly to an eye that is deliberately looking for them. Check the spacing with calipers or dividers, or by any other means you wish to employ. Check that all the spikes are set centrally between the edges of the pulley.

Spikes are often difficult to remove, so if one needs replacing or shifting it is best to file it off flush and drill it out. Use the side of a drill or a rattail file to enlarge the hole in the desired direction, and then round it out with a larger drill. (A punch can be inserted into the hole to knock out the spike diametrically opposite, if necessary.

Turn down a stub of $1/8''$ or $3/16''$ mild-steel rod on the lathe until it is a tight press fit for the hole just drilled, and hammer it into place. Cut the rod off, leaving it a little too long. Repeat the process as required with other spikes.

Mount the smallest pulley—invariably the strike-side one—on a firm-fitting mandrel in the lathe and machine the tips of the spikes and the link-bearing lands at their bases to a common size. Leave both diameters as large as possible, machining very little off the lowest surfaces. Machine the sides of any new oversize spikes which have been fitted, so that the spikes are effectively centered across the width of the pulley. The lands on both sides of the spikes must be of equal height.

Now turn down the larger pulley—the time-side—to the same dimensions as those just machined for the strike-side pulley; it may be possible to use the T1 arbor to do this, with a live center in the tailstock for stability.

For both pulleys, file the tips of the spikes to uniformity wherever required. Machining the lands will naturally have lengthened them, so file out the indents between them (Fig. 75c), making all the indents as uniform as possible. Keep trying the chain until it fits snugly round the pulley. Despite the small reduction in diameter due to machining, it is unlikely that the spikes will be too long, as they were probably on the short side initially. However, if the chain is in any way reluctant to come off the pulley, the spikes must be slimmed at their tips or even shortened very slightly, but left as long as possible.

If there is no chain available that will fit the remachined pulleys properly, it may be necessary to make one in order to complete the repair. In that case, see under **Chainmaker** above.

As for wear in other parts, the counterweight usually wears away in one place rather than equally all around. If the groove is unacceptably deep, it is best to cut through it and join the ring again. Since the ring is generally made of lead, it will soft solder readily. The reduced size will not present a problem, and the counterweight will perform as before.

The weight pulley may require attention; if it fails to keep the chain in alignment, a twist may develop. Fig. 75d shows a cross section of a chain pulley. The two bearing surfaces, or lands, must take all the weight, so it is essential that the clearance groove be deep enough to enable them to do so. The groove must also be wide enough to avoid jamming the chain.

Fig. 75e shows a cross section of the common type of pulley used in a wide variety of applications. When cable rather than chain is used in a Huygens system, both the weight pulley and the spiked pulleys have this configuration. When chain is used, the type shown in Fig. 75e is a poor choice for the weight pulley, because the links rotate at random when passing over the pulley; this promotes the development of twists. In contrast, the pulley of Fig. 75d tends to keep the links in line with those on the spiked pulleys.

In all Huygens systems where a chain is used it is essential to examine every single link and correct any that are malformed. Clean the chain and fit it on the pulleys, and take care not to include a built-in twist when joining the ends. Do not oil it.

Key shrinker: A well-fitting clock key is not generally considered to be of much importance, yet continued use of a loose key will slowly ruin the winding squares and thus create a severe problem. In time, wear makes a key somewhat trumpet-mouthed, but this can be corrected so that the key again fits snugly.

Fig. 76 shows a quickly made, easy-to-use key shrinker which will save many a key from being discarded. It is simply a ½″-long section of 1″-diameter mild-steel rod with four no. 10-32 NF tapped holes in it. The key is inserted in the central hole and the screws turned individually until the key is again a good fit on the winding square.

A key can be enlarged by driving a suitable square steel drift into it, and then hammering the outside gently all around it until the drift loosens. Neither operation takes long, but their use revitalizes many otherwise useless keys.

Let-down key: a device used to let down (i.e., unwind) mainsprings prior to dismantling a movement (see page 45).

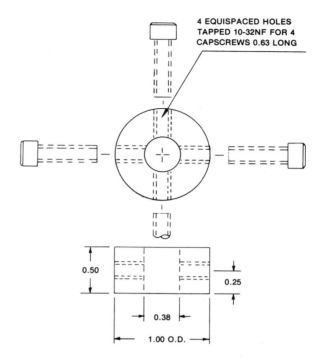

Fig. 76. Key Shrinker

Lifters: Also called lifting pieces, these are the linkages that couple the time train to the strike train. (See Fig. 7.)

Longcase: This clock, also known as a grandfather, is a free-standing floor clock usually 6′–8′ tall.

Mainspring: a spring which provides power for a clock train.

Mainspring clamp: a C-shaped steel clamp used to retrain a mainspring.

Mainspring: *Fitting.* When fitting a new mainspring, the dimensions of the new spring must closely match the old. This particularly applies to the thickness, as this has a much greater influence on the strength of a spring than the width. In measuring the old spring's thickness, make sure that it is clean and pulled out straight at the measuring point, or the micrometer will read high, making the spring appear to be thicker than it really is.

Even in open mainsprings, the width must fit the space available; however, some variation in length is often permissible. For example, an American-type movement in good condition will run for about twelve days, so if a 108″-long spring is substituted for a 120″ spring, no change will become apparent, since the clock is wound weekly in any case.

Make sure that the spring hook on the winding arbor is firmly set and strong enough to hold the spring; it must be undercut, so that the spring cannot slip off. Carefully bend the innermost coil, so that it fits snugly around the arbor and hook. It is most exasperating to have to take a movement apart again after assembly, simply because the mainspring became unhooked.

Mainspring in barrel: Encasing mainsprings in barrels has long been a tradition of European clockmakers; the idea found little favor in America, where it was undoubtedly looked upon as expensive and unnecessary. Functionally, this is undeniably true; however, a barrel does make for a neater movement. It also keeps the spring cleaner; also, spring grease does not dry up as quickly in an enclosed space. Its use is a typical manifestation of the different philosophies explained in Chapter 5, "English Clocks."

Due to space restrictions, when renewing a spring in a barrel dimensional options are almost nil. Basically, the inside coil of an unwound spring must occupy the same position as that used by the outside coil of the spring when wound. This allows the barrel to give the optimum number of working turns, which is determined by subtracting the number of coils of the unwound spring from those of the wound spring.

Regardless of whether it is in a barrel or not, a sticky mainspring is unable to supply enough power to drive an otherwise clean movement, because the coils can no longer slide over each other. During an overhaul, therefore, mainsprings must be given the full treatment, lest all other work be in vain. Once this is realized, it becomes easier to do a job that if neglected amounts to a standing invitation to trouble.

There are several methods of removing and replacing mainsprings in barrels. (See also **Mainspring winders** below.) Despite a lack of total agreement on this question, I consider it best to do both jobs by hand. Spring distortion, which is claimed by opponents of this method, is of rare occurrence and negligible effect, if subsequent performance is any guide. Fusee springs are admittedly hard on the fingers, but they are not encountered often enough to distress the average clock addict.

Let us proceed by hand then (Step 13). Check the barrel arbor for end play, which should not be more than about 0.015″—0.010″ is better. After some experience, such clearances can be checked by feel; they need not be measured every time with a dial gauge. Any excessive end play will be taken up later, by slightly dishing the end of the barrel, as shown in Fig. 77. If the barrel or its cover obviously needs bushing, end play can be adjusted by the new bushing (as explained above under **Bushing: Barrels**).

The barrel cover is a press fit into a recess in the rim of the barrel, and a slot is provided for its removal. Using this slot, pry off the barrel cover with a tapered awl or suitable small screwdriver. Pull up a few coils of spring from the center, and check their condition. They will generally be either dry and dirty, or very oily and dirty—the latter condition being due to some enthusiast's squirting household oil into the barrel in an attempt to make the clock run. In either case the spring must be taken out and thoroughly cleaned and greased, for household oil dries up relatively quickly, thereby adding to the black, sticky deposits already present.

It must be emphasized here that barrels with springs inside them must never be put into the **cleaning machine,** or into liquid of any kind, for that matter. No solvent can penetrate to any useful effect between the closely packed coils of the mainspring, so they will remain dirty and sticky. Also, the barrel does not drain easily, and fluids cannot circulate freely inside it, so any attempt at rinsing will fail. The spring must therefore be removed from the barrel before both are cleaned and rinsed.

Most mainsprings are not too hard to control as they come out, but if you are in any doubt, and a powerful spring is involved, surround the barrel with a thick cloth such as an old towel before pulling up the center coils and letting the spring come out under its own power. A dry mainspring is less likely to come out in a hurry, but in any case spreading your curved fingers over the towel will keep things more or less under control. If, when the spring has emerged, the end is still held by the barrel hook, do not use force to free it, for this will only twist the end of the spring. Instead, push the end of the spring firmly back into the barrel; the spring will then come free from the hook.

The end of a mainspring generally leaves a mark, outlined in oily dirt, on the inside of the barrel. Whether this is present or not, it is a very good idea to scribe such a mark immediately, as a reminder of the sense in which the spring is wound—that is, clockwise or counterclockwise. It can be a bit depressing to wrestle a powerful spring back home, only to find that it is lying the wrong way round, and that it all has to be done again.

Unless it is obviously cracked or damaged, the mainspring now goes into the cleaning machine, along with everything else (Step 15); this includes the barrel, its cover, and its arbor. After cleaning and drying all the parts, carefully check the mainspring again for cracks or rust. Cracks usually mean a new spring is required; rust can be removed with emery cloth. If the barrel has any broken teeth, or a loose or protruding spring hook, see **Tooth Replacement: Going barrel** below.

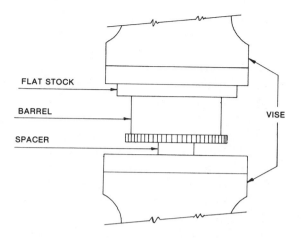

Fig. 77. Dishing a Barrel

Now set the arbor up in the lathe, and polish its bearing surfaces. In most cases, there are four such surfaces: two on which the barrel rotates, and two normal arbor pivots. Note that aside from the initial setting up the arbor pivots of a fusee-spring barrel do not turn; hence, these particular pivots do not require much polishing; it is sufficient that they be clean and a good fit in their pivot holes.

Assuming that barrel bearings and arbor end play are satisfactory, the mainspring should now be greased and put back into the barrel. Apply the grease with your fingers, and ensure a thin, even layer over the full length of the spring. A little clock oil is acceptable for the center coils, if required.

It is not usually necessary to secure a barrel while feeding a spring into it, but for those who feel happier with the barrel held down, a spotface in a block of wood works well. This recess is easily made in a short end of $2'' \times 4''$ pine, by using an expansive bit in a woodworker's brace. Make the spotface deep enough to cover the teeth of the barrel—say, $3/16''$—the spotface diameter must assure a light press fit for the teeth.

Check that there are no tears around the hole in the end of the spring; if even a small tear is present, refer to **Mainsprings: Repair,** below. Never put back a spring that has small tears or cracks around its retaining hole, for they will lengthen until the end of the spring breaks off, causing considerable damage.

To return the spring into the barrel, bend its outer end into a curve approximately that of the outside of the barrel, and see that the tip of the spring does not prevent the hook from holding it securely. Assuming a clockwise feed-in, and with the barrel hook furthest away from you, say at "twelve o'clock," bend the end of the spring, including the hard section, into a J-shape over some 200° of arc, and introduce this into the barrel with the hole at about "eleven o'clock." There will be a strong tendency for the end of the spring to creep around the inside of the barrel in the clockwise direction. Allow it to do so slowly, until the hole clicks over the hook and the spring is held behind it. Then, holding the spring with your left hand, push the end of the spring behind the rim of the hook with a screwdriver, if it does not click in on its own.

Now feed the rest of the spring into the barrel, taking care to see that the first coil does not foul the top of the hook projecting through the end of the spring. Do not stop until all of the spring is home. A few taps on a block of wood with the barrel open end up will bed the spring coils down evenly in the barrel.

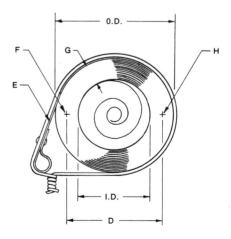

Fig. 78. Measuring a Coiled Mainspring

Adjust the inner coils of the mainspring to a good fit with the arbor, and set it in place. Replace the barrel cover by simply pressing it into place in the vise. Some covers have a tapered edge on the outer side or a projecting bushing on the inner side. Some have both. After being replaced, the cover should be tapped down gently on the open vise with a hammer, using a piece of ½"-birch dowel as a flat punch. If the cover is slightly loose, beat it out a little by hammering it evenly around its edge, taking care to keep it round and flat and not to expand it too much.

Wipe the barrel clean of spring grease with a rag and a little rinse solution; do not use a general-purpose solvent, as it leaves a sticky residue.

Mainspring: *Measuring.* Because the dimensions of a coiled mainspring are not always known, it is useful to know how to make a fairly accurate estimate of them yourself. Fig. 78 shows a typical new, wire-held mainspring; a used mainspring would have more free coils in the center space.

The width (not shown) is simply measured with a micrometer, and the thickness can often be measured likewise at E in Fig. 78. If E is inaccessible, measure G with a vernier caliper instead, and count the number of coils across at the same point. Dividing the G dimension (in thousandths of an inch) by the number of coils will give the spring thickness; accuracy here requires that all the coils be touching each other. To count the coils, slowly drag a toothpick or pegwood stick across them and count the clicks; wood is preferable to steel for this purpose.

To obtain the length, first find the average diameter of the close-packed turns. This is equal to half the sum of the O.D. plus the I.D., as measured. A simple and quick method is to lay the points of a vernier caliper across points F and H by eye; with practice you can use this technique quite accurately.

Having found the diameter D of the average

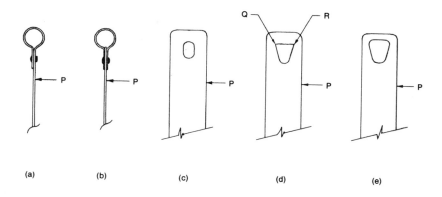

Fig. 79. Mainspring Repair

turn, the length L of the compacted turns is

$$L = \pi DN$$

where L = length in inches
$\pi \cong 3.14$
D = average diameter in inches
N = the number of coils

Measure the lengths of the extended loop end and the inside turns; this last can be done with dividers, or by counting the number of estimated inches with a pencil. The inside turns on a new spring may be only 4″, while those of a used spring can take up 1′ or more. Adding these two dimensions to L will give the total length of the spring to within a few inches; this can be verified by practicing with a spring of known length.

Mainspring: *Repair.* In general, only the outer end of a mainspring can be successfully repaired, or rather, shortened. Repairs to the inner end often fail; in fact, if more than the innermost turn is removed, the spring usually breaks at the first winding. This is because the tempered inner coils are unable to bend around a reduced radius, and as more is cut off, the worse the problem becomes.

A repairable spring is one which has lost no more than some 7 percent of its total length, measured from its outer end. If you want to make a loop, as in Fig. 79a, soften about 1½″ of the broken end by heating it red hot and allowing it to cool slowly. (Point P throughout indicates the limit of softening.) Now cut off the ragged end and round the corners with a file. Form a new loop by bending the end round a rod of appropriate clock-pillar diameter, and drill and rivet it as shown (Fig. 79a).

If the old, separate loop is available, soften only 0.8″ or so and drill and rivet (Fig. 79b).

For barrel-enclosed springs, soften about 1″ of spring, as before, and clean up the ragged edge. Reshape the end, and drill and file a hole as shown (Fig. 79c). Elongate the hole with a round file so that it will accept the barrel hook comfortably; do not leave sharp corners and do not make the hole unduly large, as this only weakens the repaired area.

The hole shown in Fig. 79d is unacceptable; it is likely to tear at the sharp corners Q and R. The hole at Fig. 79e is too large and too close to the end of the spring; both 79d and 79e must be cut off and remade to the pattern of 79c.

Holes can be punched through tempered-steel spring, as described under **Pendulums,** but there is some risk of cracking here; the softening process is safer for mainsprings.

Mainspring winders: Those commercially available seem to fall into two classes—very dubious and very expensive. One proof of ownership of the first type is lacerated hands; of the second type, loss of the price of a pretty good clock. Collectively, I trust them not.

Open mainsprings can be wound up using the movement itself (Fig. 16). Enclosed mainsprings require the expensive type of winder; unless a considerable volume of work is involved, the expenditure of several hundred dollars is hard to justify. It is feasible to use a lathe as a winder, but an appreciable amount of setting up is required, which includes machining quite a number of components to suit different sizes of springs and barrels.

Although other techniques sometimes come to mind, and a winder is an interesting device to think about, I still prefer to use my hands.

Maintaining power: This is the force which keeps a clock running during winding. When incorporated, the mechanism is found chiefly on the better grades of weight clocks, such as Viennas, where the driving force is removed during winding. Clocks with going barrels have maintaining power built in, because the driving force actually increases during winding.

Maintaining power is most commonly applied by John Harrison's device of driving the train via a spring-loaded ratchet gear, which couples the winding drum to the T1 gear instead of the usual direct drive.

Without maintaining power, there is a slight chance of damage to the escapement due to the butting and checking which takes place during winding. A Vienna regulator may develop a temporary pendulum wobble. A long-case clock with a recoil escapement may well run backwards during the ten seconds it takes to wind it up. This means that the winding time must be taken into account; it must actually be regulated to run twenty seconds fast, and the time taken to wind it each week should not vary.

The maintaining spring must be weaker than the driving force, but still strong enough to drive the train.

Meshing: A gear and pinion which are meshed too lightly—that is, too far apart—cannot transfer adequate energy, for the teeth tend to butt against the pinion leaves or trundles, instead of driving them round. On the other hand, when the meshing is too deep the tips of the gear teeth may bottom in a cut pinion, or hit the arbor in a lantern pinion; optimum meshing is therefore essential. (See Fig. 6, and above under **Gear expander**.)

Oiling: Clock movements should be lightly oiled in certain places, using a good grade of fresh clock oil. Nowadays organic oils appear to be losing ground to synthetic oils, and these appear to be quite as beneficial for clocks. (The use of synthetic oils is even more beneficial to the creatures who pay so dearly in supplying the other oils.)

Light affects many oils adversely, so it is best to keep clock oil in a dark or shaded container. Since most oil bottles readily overturn, a wooden holder that both shades and holds the bottle at 45° is a most useful device to make.

A round bottle is easy to accommodate by drilling a suitable hole with a wood bit; square bottles can be set up by gluing flats and blocks together. Sections of old mainspring can serve as retainers, and a shelf for an oiler can also be incorporated.

An oiler need only be a few inches of 0.032″-diameter steel wire, with an inch or so of ⅜″-diameter wooden dowel as a handle; the tip of the wire may be advantageously flattened and spear-pointed.

Dip the tip of the wire oiler into the oil, touching it on the bottleneck as it comes out, so as to leave only a tiny droplet on the tip. Apply one such droplet to each pivot hole of the clock; it is best to oil one plate at a time, working from the top downwards. Take care not to miss a hole that may be hidden behind a lifter. Owing to its large size, slow speed, and light load, there is no need to oil the hour pipe internally; only where it passes through the front plate is lubrication required. In American clocks, oil both sides of the count-gear bearing.

Oil both sides of the saddle, the groove in the fly arbor, the clicks and ratchet gears, and the barrel bearings. Apply a slightly bigger drop of oil to the largest bearings. If any oil runs down the clock plates, you are applying too much. Do not oil any other gears or pinions; this does more harm than good, because wear will result when the oil collects dust.

All oils slowly deteriorate with age. For oil in a pivot hole, a life of five years is about average. Do not try to economize by using old oil, for it will dry up sooner than expected, and the clock will require cleaning again; always use fresh oil.

Pallets: Those surfaces of the verge or anchor which engage the teeth of an escape gear.

Pegging out: Several sizes of pegwood are available, often in bundles of about fifteen. The sticks are usually about 6″ long, and range from 0.120″ to 0.230″ in diameter. The wood is firm but can be sharpened with a pocket knife, pencil-fashion, to a tapering point.

To clean out the pivot holes in a clock plate, push the point of the pegwood into each hole in turn and rotate it until the hole is clean; then repeat this from the other side of the plate. The wood will require sharpening frequently as it breaks or becomes dirty and scuffed. If other wood is required for the larger holes, use a dry hardwood, never a wood with any suggestion of resin in it.

Pendulum cock: the support from which a pendulum hangs.

Pendulum: When linked to a clock, a pendulum becomes a swinging governor which de-

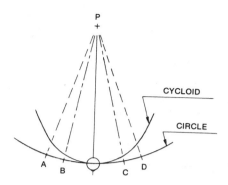

Fig. 80. Pendulum Arcs

termines how fast the clock will run. The clock supplies enough in-phase energy to overcome normal frictional and air resistance losses which would otherwise bring the pendulum to a standstill. To a large extent, the balance between these losses and the energy supplied determines the length of the arc.

Unfortunately, a pendulum is not the simple weight-on-a-wire device which at first it might appear to be. Fortunately, however, most of us do not have to dig too deeply into its complications to use it intelligently, and the basic practical details are not too hard to master.

Fig. 80 represents a typical pendulum, hung by means of a suspension spring at P. Over the arc range in question, the bob can be assumed to swing in a circular path. With this configuration the bob takes a little longer to cover the long arc AD than to cover the short arc BC. If, however, the bob could be persuaded to take a cycloidal path, then dissimilar lengths of arc would take the same time. Just who it was that first worked all this out is open to argument, but the credit generally goes to Christiaan Huygens of Holland. The phenomenon is called **Circular error.**

However, all attempts to achieve a cycloidal path for the bob have failed; subsequent efforts went into trying to keep the arc constant. To most people this seems to be synonymous with keeping the pendulum as detached from the movement as possible; however, if the coupling between them is constant, good results must also follow. Naturally, the impulses supplied via that coupling must also be constant.

In the early eighteenth century George Graham noticed that the circle and the cycloid are very close together for about 3° each side of center. He then proceeded, around 1730, to design his successful deadbeat escapement to take advantage of that fact.

Note that in practice variations in the pressure applied to the train (e.g., by a mainspring) far outweigh circular error considerations as far as timekeeping is concerned.

A full dissertation on pendulums can easily occupy a book in itself, so I shall limit myself to practical information and simplified methods of using it.

A theoretical pendulum is a point-size weight swinging on a weightless wire. Even though such an abstract device is not exactly common, the concept does enable us to calculate the lengths of pendulums. The point is known as the center of oscillation; in practice, it is situated a little above the middle of the bob; this is because the pendulum shaft, be it wire or wood, does weigh something and must be taken into account. A pendulum also has a center of gravity and a center of gyration, but only the center of oscillation need concern us here.

Because the acceleration due to gravity varies slightly over our planet, pendulums, being gravity operated, also must vary a little in length from place to place if they are to beat at the same rate. In the calculations that follow, I have taken G (the acceleration due to gravity) as being 32.16 ft./sec./sec. In speaking of vibrations or beats-per-second, most writers fail to state whether they mean a single swing from A to D (Fig. 80) or a swing from A to D and back again; unless this is made clear, confusion results. Another important point—evidently not apparent to beginners—is that to make a given pendulum beat twice as fast, its length must be divided by four, not two. (As a boy, I found that out the hard way!)

Most clock enthusiasts have, hung on a nail or in a box, one or more movements of which the pendulum lengths are unknown and would be time-consuming to determine. Usually, only awkward attempts at gear and pinion counts and beats-per-minute tables can be made, and the whole process is half forgotten by the time the question arises again. Bearing in mind that one beat—one one-way pendulum swing—releases half a gear-tooth space, a simple movement-related formula can be worked out as follows:

Let us say that the escape gear has N teeth and revolves R times per hour; then we can state that at any given point $N \times R$ teeth pass per hour (3600) seconds. Therefore one tooth passes in $\frac{(3600)}{NR}$ seconds, and the time t for one beat (which represents half a tooth space) takes half as long.

Hence,

$$t = \frac{1800}{NR} \text{ seconds} \qquad (1)$$

In terms of L and G,

$$t = \pi\sqrt{\frac{L}{12G}} \qquad (2)$$

where L = pendulum length in inches,
G = acceleration due to gravity
= 32.16 ft./sec./sec.
$\pi \cong 3.14$

(These units also apply to equations (5) and (6) which appear later in this section.) Equating (1) and (2), we have

$$\frac{1800}{NR} = \pi\sqrt{\frac{L}{12G}} \qquad (3)$$

In round figures this simplifies to

$$L = \left(\frac{11260}{NR}\right)^2 \qquad (4)$$

Set the movement on the bench; put a pencil mark between two escape-gear teeth, and count N. Then, to find R, remove the verge, and firmly attach a longish, spare minute hand. Applying both forefingers to the time-train gears, bring the hand up to a quarter-hour mark—say, the 45-minute mark—and note where the pencil mark has stopped. Avoiding backlash, "finger walk" the gears so as to allow the hand to move forwards fifteen minutes, counting the escape-gear revolutions meanwhile. Multiplying this number by 4 gives a reasonably close value of R; greater accuracy is obtained by counting for a full turn of sixty minutes.

Using these figures for N and R in equation (4) gives L, the theoretical pendulum length. In practice, this is measured from the bottom of the pendulum cock to a point slightly above the center of the bob; the amount above the bob center varies with the type of pendulum. Only a heavy shaft or a bob with uneven vertical weight distribution will markedly upset this approximation. It is wise to leave the shaft a little on the long side of your calculated figure, since shortening it is easier than lengthening it. A typical allowance for the usual wire or wire-and-wood, 15″ school-clock pendulum would be ½″.

There is no need for great accuracy in working out practical pendulum lengths, because the center of oscillation is not directly measurable on the bob itself; the final length must always be found by mechanical adjustment.

As equation (4) relates the number of escape-gear teeth directly to the pendulum length it is very convenient to use. All constants (π, L, 12 and G) are included in a single number, and it is not worth changing it to accommodate the small differences which occur in G values. By examining your clock movement and using the above formula, it only takes about ten minutes to determine the required pendulum length. The effect of changing the number of teeth on the escape gear can also be quickly calculated.

Another useful equation which relates L and t is

$$L = 39.14 t^2 \qquad (5)$$

Table 6. Pendulum Data

S	L	S	L	S	L
60	39.14	100	14.09	140	7.19
65	33.35	105	12.73	145	6.70
70	28.75	110	11.64	150	6.26
75	25.05	115	10.65	155	5.86
80	22.01	120	9.78	160	5.50
85	19.50	125	9.02	170	4.88
90	17.39	130	8.34	175	4.60
95	15.61	135	7.73	180	4.35

Typical pendulum rates and lengths are given in Table 6, where

S = number of one-way swings per minute.

The equation relating L and S is

$$L = \left(\frac{375.36}{S}\right)^2 \qquad (6)$$

One occasion for rebuilding a pendulum is when matching disparate clock cases and movements. If you keep an empty clock case long enough, a movement may come to hand—although too often it will be a week after you traded the case! Sometimes the movement, although of the correct make, turns out to be for a case somewhat longer than yours. The only thing to do is shorten the pendulum, and this means changing the time-train gearing; usually it is enough to exchange the escape gear for one with more teeth and refit the verge to suit.

Let us suppose that we wish to take 2″ off a 13″ pendulum length L_1, to make length $L_2 = 11″$, and our new escape gear has N_2 teeth. To find N_2, we can use equation (6), which gives us

$$\sqrt{L} = \frac{375.36}{S}$$

Hence,

190 GLOSSARY

$$S_1 = \frac{375.36}{\sqrt{L_1}} = \frac{375.36}{\sqrt{13}}$$

= 104 one-way swings/min.

and

$$S_2 = \frac{375.36}{\sqrt{L_2}} = \frac{375.36}{\sqrt{11}}$$

= 113 one-way swings/min.

In our example, the old escape gear has thirty-four teeth, and, since both old and new escape gears must run at the same speed R (revolutions per hour), let us use formulae (4) and (6) to express R:

$$\left(\frac{375.36}{S}\right)^2 = L = \left(\frac{11260}{NR}\right)^2 \quad (4) \text{ and } (6)$$

$$\therefore \frac{375.36}{S} = \frac{11260}{NR}$$

which simplifies to

$$R = \frac{30S}{N}$$

$$\therefore R = \frac{30S_1}{N_1} = \frac{30S_2}{N_2}$$

Substituting our known values gives

$$\frac{30 \times 104}{34} = \frac{30 \times 113}{N_2}$$

from which we derive

$$N_2 = 36.94$$

Hence an escape gear of thirty-seven teeth would match our new 11″ pendulum.

Pendulum

Practical pendulums come in many shapes and sizes, but they all work in the same manner. The simple wire-and-bob American type has almost all of the weight concentrated in its bob, and so comes close to the theoretical pendulum. Another American pendulum, found on long-drop school clocks, has a joint about halfway along its length to permit removal of the bob and its lower wooden shaft. Since this joint can cause a loss of drive, it is often made into a dual joint to obviate flexing. When checking this type of pendulum, always sight along its two sections, and adjust the two hooks of the joint until the whole shaft is straight; this equalizes the load on each side, and thus allows the joint to perform as intended.

Make sure that the rivets or screws are holding the hooks firmly in the wood, and that the rating screw is also firm. If splits are evident, either part can be refitted in a hacksaw slot, using epoxy resin. Clamp the joint while the resin sets. Wood screws and rivets in wood tend to loosen in time, and it is better to use epoxy, and then follow it up with a pair of 2-56 NC screws with washers and nuts. This prevents the wood from warping and opening up the joint at some future date. Similar joints are sometimes found in German clocks.

The English longcase pendulum—including the method of setting it up—has been dealt with in Step 49 of the longcase 8-day weight movement (page 82). As these 1-second pendulums are quite heavy, if one is dropped, the rating screw often bends or breaks off; the suspension spring also is easily damaged. These

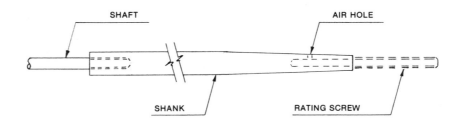

Fig. 81. Longcase Pendulum Repair

and other pendulums should therefore be checked before installation.

If a bent rating screw cannot be straightened, it must be replaced. In most longcases, you will see that the screw extends from the bottom of the heavy steel shank on which the bob slides. Saw the defective screw off, and drill out the stub—preferably with a no. 29 drill—so that it can be tapped 8-32 NC. Keep the drill in the center of the shank; however, it does not matter too much if the tap shows signs of breaking out, provided it does so equally on both sides. If the shank is too thin for tapping, simply turn off the screw threads, on the lathe, for about 0.60", drill a clearance hole to suit, and solder the screw in place (see Fig. 81). To avoid bubbling, drill a tiny air hole, as shown, and rotate the screw while the solder is molten; used with flux, this ensures good tinning and a solid joint. The same method can be used at the other end of the shank, if the shaft happens to be loose, and again where the shaft enters the block at its upper end.

The bob must slide on the shank without being loose enough to wobble. Play can be taken up by gently hammering the lead or using a strip of shim if space permits. Look down the length of the shaft and straighten it where necessary. Make sure that the rating nut is large enough to make a precise contact with the bottom of the bob; a small nut may make an uncertain contact, and so cause adjustments to be erratic. As explained in Step 49 (p. 82), the cause of wobbling is almost always at the top of the pendulum.

French clocks are mostly of the mantel variety, and so have short pendulums; unless one side of the customary double suspension spring is broken, they do not give much trouble, except from the accessibility viewpoint.

German spring-driven wall clocks often have imitation temperature-compensated pendulums with brass and steel rods alongside the pendulum shaft; these "R & A" ("retard-and-advance") pendulums, as they are called, are decorative, but they do not compensate. Sometime the rods are all of steel, with some plated to look like brass. (Genuine temperature-compensated pendulums are only found on clocks of very high quality.)

The shaft joint on these clocks is only a few inches down from the suspension spring, at the bottom of the pendulum hanger, as it is called. Some form of non-flexing joint is used, and the two sections of the shaft should be brought into line, for the reasons previously given. Make sure that the rods penetrate the bob and thus steady it, and that they cannot slide on the central shaft and so upset the timing. The positioning of these rods is quite significant.

The dual, side-by-side suspension springs can be repaired when broken, provided that brass chops are used, but it is much easier to replace them. Viennas use the same arrangement, and their full-length wooden shafts are treated in the same way as those of American long-drop school clocks.

A simple out-of-sight pendulum bob for mantel clocks can be turned up quite quickly on the lathe. Cut off a ¾" length of ¾"-diameter mild-steel rod and face off both ends. Mount the ¾" length of rod in the lathe and drill a central hole through the rod just large enough to accept the threaded section of a bicycle spoke; note that the rolled threads require a larger clearance hole than the spoke-shaft itself. Use a piece of copper or other wire as a temporary shaft to find the required length, and then replace it with the spoke, after cutting it to size and bending a small hook on it. This must be done with the nipple set in the middle of the threads, so that adjustment is possible.

Other dimensions of bob can be used, depending upon the required weight; such utility pendulums will serve well where appearance is not important.

When forming pendulum hooks, including that described above, always use the form shown in Fig. 82f. Although the hooks shown in 82a, b and c are common, they will only work well when used with a wire loop as in 82b. The kink caused by the kitchen-clock pendulum in 82e cannot occur with the 82f configuration, which is excellent for all types of loop.

The flattened tops of the oval rods used in kitchen-clock pendulums should be filed to the dotted line, as shown in Fig. 82d; this helps prevent kinking and also improves the appearance of the loop. Make sure that the oval hole in Fig. 82d is a free and unambiguous fit at T for the hook, and that there are no sharp edges on it.

The vertical creases sometimes found in the tops of these loops are a poor idea, and it is best to flatten them out and tidy up the loop as mentioned above. To have any influence at all, these creases must inevitably kink the pendulum shaft along the lines of 82e, and such kinks, being unstable, are a likely source of wobble.

In American mantel clocks, wobbling can also occur if the timing mechanism is loose or its slots are too wide to fit the suspension spring closely. The flat section of the suspension spring must always be in line with the wire. Check this by sighting along the wire, and adjust the top of it accordingly; if the edges of the suspension spring are curved, they must be trimmed straight with shears. The pendulum and its suspension are far more important than is generally realized.

Philosophy of clock repair: As your interest in clocks becomes known, your assistance will occasionally be requested by friends and acquaintances; you will then have to balance quality of workmanship against time spent, financial feasibility, doing a favor, and so on.

You may be introduced to a man who clutches a horological ruin (known as a "basket case" or "dog") that "needs a little work"—such as an entirely new movement and case. It seems that he wants it fixed—for not more than ten dollars—so that he can sell it. Don't assume that he is a sharp operator; more often than not, some hard citizen caught him when his guard was down, and now he just wants his money back.

If you attempt the impossible, you will fail, and come to regret it. Remember that your customer will not tell a prospective buyer, "I just had this patched up to go a bit, so I could sell it." No way: He will say, "I've had this clock completely overhauled by so-and-so"—your charming self. Now unless you have done a proper overhaul, whoever winds up owning it is going to discover just how horrible the clock's condition really is, and if your name is associated with it you will not hear the last of it for many a long year.

It matters nothing to a clock whether it is cherished or thrown away; if certain things must be done to make it run, then you either do them and it runs—or you don't and it

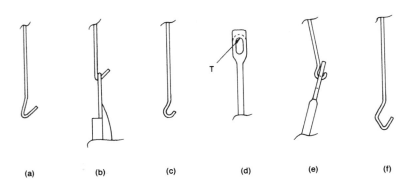

Fig. 82. Pendulum Hooks

Pinions

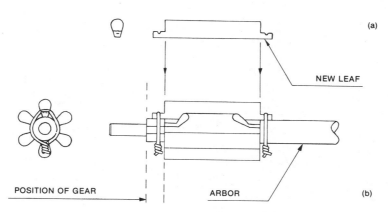

Fig. 83. Fitting a New Pinion Leaf

doesn't. A customer needs to have this explained to him—particularly the man with the ruin—so that he understands why you cannot, on his terms, touch his job with a sterilized pole. Basically, if it takes twenty-six hours to effect a proper repair, then twenty-six hours have to be put into it; you cannot argue with a clock, but you *can* talk about time and money with its owner. You will get quite enough no-win jobs, which take far too long, as it is, without undertaking them knowingly.

You will have to set a value on your time and expense, for clock parts are not cheap; tools wear out, and oil, grease, and electricity all cost money. You should give a guarantee; you cannot guarantee a spring, but other work should last for several years, and it is not in your interest to charge twice for the same job. Guarantee it for a year, and accept that you have some obligation even after that.

Always do one thing well before passing on to another, and do not differentiate between a job for yourself and a job for someone else. You cannot switch on and off in the struggle for perfection; only continual effort will bring you even close to it. And remember that price is not the same as value: A job has a price, but your reputation has a value; being priceless, it is worth more than all the jobs you will ever do.

Pinions: These are really small gears; in clocks they are either cut pinions or lantern pinions, both types being driven for the most part by gears much larger than themselves. Thus driven, lantern pinions are the better of the two, as they have less friction and can easily be repaired when worn.

Pinions: *Cut.* These pinions are integral with their arbors; when worn out, their replacement generally requires a new arbor to be made. Sometimes it may be possible to turn the old arbor down and silver solder a new cut pinion onto it. A suitable hole must be drilled through the new pinion; however, obtaining enough strength at the pinion may be a problem.

The old cut pinions of longcase clocks are often deeply undercut, and, being thus weakened, may have a leaf break off; this occasionally happens on the minute arbor, T2.

To fit a new leaf, any brass gear present must be removed; because it must be replaced exactly as before to avoid eccentricity. First scribe a line on it opposite the broken leaf.

With the gear off, file the remains of the old leaf flat, and clean up the surrounding metal. File out a new leaf from a piece of steel as shown in Fig. 83a, leaving it a little on the high side. Wire the new leaf in position, as shown in

Fig. 83b, and set it up truly; it helps to crimp the wire loops so that they lie close alongside the new leaf. When all is set up, apply flux (borax), and then silver solder the leaf in place. Remove the wire loops and flux residues, and file the ends to size.

It is not easy to obtain a tempered leaf, because quenching a silver-soldered joint may weaken it, and even if the new leaf could be tempered it would be too hard to trim with a file. Tool steel in the annealed state will hold its own with the remaining, worn pinion leaves, which may not have been tempered originally.

Pinions: *Lantern.* These pinions consist of several slim rods, known as trundles, held in position between two collets or shrouds (see Fig. 4). Repairing lantern pinions is relatively easy; you simply replace any trundles that are worn or bent. Dirt tends to fall through a lantern pinion, whereas with a cut pinion it is compressed between the leaves.

The two faults generally exhibited by lantern pinions are bent, grooved, or missing trundles, and loose shrouds. A defective trundle can be extracted by forcing it up through the shroud, using a small pair of long-nosed pliers. Small flats may be ground on the outside of the plier tips for this task, to ensure entry between adjacent trundles. No great force is required to extract the trundles, for they are held in place only by light indentations in the shroud. Retain a relatively straight, unmarked specimen, and measure its diameter as a reference.

If no suitable thin steel rods (pinion wires) are available, then good-quality, round, highly polished sewing needles of hard steel make excellent replacements. In fact, they are even better than the originals, being smoother and harder. Use a micrometer to select a needle of suitable diameter, and break off the tapering point of the needle. The rest of the needle—that is, the parallel section—is now inserted into the trundle hole as far down as it will go; it is then broken off as close to the shroud as possible.

When all the trundles have been inserted, they are easily held in by soft soldering, which will also ensure that the shroud does not turn on the arbor. A method that is sometimes tried—and proves far from easy—is to work the loose shroud off its arbor, shrink the hole by hammering it up, and then drive it back on again amid free-milling trundles. However, there is no certainty of a tight fit by this method, and I do not consider it to be worthwhile.

If the trundles are only slightly grooved, and therefore still strong, they can be rotated until a new, smooth surface faces the gear teeth that drive them. If the shroud is then fluxed and soft soldered, enough solder will enter the trundle holes to hold them in this position.

A well-soldered shroud appears as a neat, low cone of bright solder, without holes, ash, or flux residues. A good way of attaining this is to clean and flux both arbor and shroud, and then rotate them in a hole in a wooden block while applying the solder via a small iron in the usual manner.

Another good device for retaining trundles that is occasionally found is a small brass disc which is driven onto the arbor until it touches the shroud. Being of the same diameter as the shroud, it makes a tidy job, but it has not been widely used, because stamping the shroud is undoubtedly cheaper.

Pivots The turned-down ends of clock arbors are known as pivots; together with the holes in the clock plates, they form rather loose sleeve bearings. Because only weak forces are involved, even a small flaw in a pivot can stop a clock train.

Fig. 84h shows a sound pivot; all the rest show typical defects. The scoring in 84a is generally due to the exertions of a center-punch artist; if it is not too deep it can be removed with a polishing stick as shown in Fig. 85a. If the scoring is deeper, it will have to be ma-

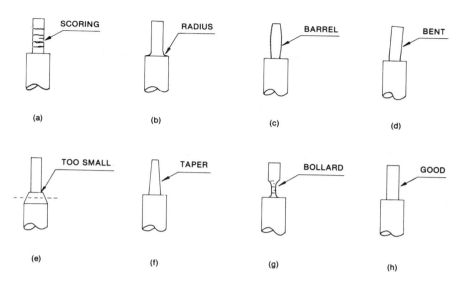

Fig. 84. Pivots

chined out—with the pivot as close to the lathe collet as possible—and then polished afterwards. The radius shown at 84b must also be machined out. Although a sharp-edged burnisher is supposed to resolve this problem, I find burnishing quite unsatisfactory, and much prefer machining.

The barrel shape of Fig. 84c can also be machined true, but, as with 84a, there must be enough metal left in the pivot to provide adequate strength; this also applies to 84f. The bent pivot at 84d can usually be straightened by means of a suitable hole in a stub of brass bushing tube, after which the stub can be chucked and the arbor rotated in it to see that its free end does not circle.

In 84e, the end bearing is too small, and must be machined back to the dotted line; this shortens the arbor and sometimes necessitates a flanged bushing to maintain correct end play.

The bollard shape at 84g is much too weak for use, as may be the pivots of 84a, c and f, if they are too thin after being turned down; all such pivots must be renewed. The 84g shape is almost always found in cheap steel-plate movements. Being basically unsound, these steel-to-steel bearings wear rapidly, and further damage is often caused when the pivot finally gives way and unmeshes the gears. If damage is widespread, as it usually is, such movements are best scrapped for parts. The only lasting cure entails extensive repivoting and fitting of brass bushings throughout, which is very seldom worthwhile. Some steel-plate movements have brass bushings in most places, but this does not make their appearance any more attractive, and rust often shows through the plating.

Repivoting generally involves drilling a hole in the end of the arbor and inserting a short length of steel rod into it. It is essential that the new pivot be located centrally in the arbor, and this sometimes poses a problem. If the arbor can be set up close to the lathe chuck, the hole can be centered and drilled, as usual, from the tailstock. However, in many cases this is not possible, and then it becomes necessary to make a steady (Figs. 88 and 89) and a chamferer (Fig. 87) in order to set up the ar-

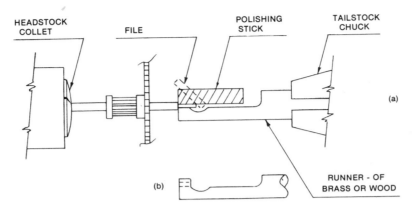

Fig. 85. Polishing a Pivot

rangement shown in Fig. 90. A good alternative is to use the **Plewes chuck** (page 200), which is designed to solve exactly this type of problem.

Take a scrap piece of ⅜"-thick mild steel, and drill a clearance hole in it to take the arbor, as shown in Fig. 86a. Push the arbor gently against a 00 file and face off the end, (Fig. 86b).

To make the chamferer of Fig. 87, true both ends of a 3" length of ¼"-diameter drill rod, and drill the two central holes as shown. Set the rod up in the vise, and, with a triangular section file, make two diametral grooves, of equal depth and at right angles to each other across each end. Keep the top of the file level, in two planes (both from end to end and across

it), and deepen the grooves until they are as wide as the rod. Check that the four cutting edges thus formed are equal and evenly spaced, and then harden and temper each end to a pale-straw.

To make the repivoting inserts, turn up two inserts—no. 1 and no. 2—from ¼"-diameter drill rod, as shown in Fig. 88. Drill the countersunk hole to the shank diameter of a center drill that has a smaller pilot diameter than that of the intended hole H. Provided that it is a good fit for the drill chosen, the size of the hole H is not critical; it depends somewhat upon the sizes of the arbors and pivots involved, as we will see. Other sizes than those shown in Fig. 88, such as 0.031" and 0.052" may also prove useful. Drill the hole H after withdrawing the center drill. The chamfer E may be put on in the lathe with a file; it need only be about 0.015" wide × 45°, but it is essential as a lead-in when press-fitting the insert. Larger inserts may also be made to accommodate thicker arbors and pivots.

The steady of Fig. 89 mounts on the cross-slide of the lathe to hold the free end of the arbor to be repivoted, as shown in Fig. 90. It is possible to make a steady for mounting in a tool post instead, but steps would have to be taken to see that it does not swivel when pushed against the end of the arbor; therefore, cross-slide mounting is undoubtedly preferable.

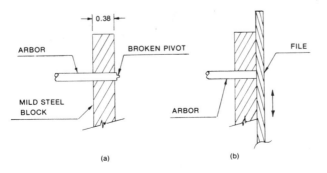

Fig. 86. Facing Off an Arbor

The dimensions given in Fig. 89, while typical, will naturally vary from lathe to lathe. The steady here is made from a piece of 2½" × 2½" mild-steel angle, ¼" thick. File both ends square and make sure that the base P is flat so that it cannot wobble when resting on a flat surface. It is preferable to use cold-rolled steel, but this is far less common than hot-rolled angle. File the scale off the front face Q, and adjust the angle if necessary in a vise, so that Q and P are at 90°.

Drill two ¼"-diameter mounting holes (R), and bolt the steady down on the lathe cross slide, lining it up at 90° to the centerline of the lathe. Because the full width of the steady is available, several holes may be drilled in it to accommodate inserts which can be used for various purposes; a typical layout is given. When dimensions have been decided, set up a center drill in the headstock chuck, and, using the cross slide, line up the first hole in the steady.

Now fit a stub of ½"-diameter steel rod (or other metal) in the tailstock chuck, and use it to drive the carriage to the left so that the center drill cuts into the steady. This method is preferable to using the carriage feed, since its direct thrust ensures accuracy by preventing the steady from leaning. Use the cross-slide micrometer dial to locate the remaining holes; this registration will be very convenient later.

Now drill a hole in a piece of scrap, ¼"-thick steel with a sharp D drill (0.246"), and check that your two repivoting inserts are a satisfactory press fit therein. If the insert is loose, use a C drill followed by a D, and try again. Then remove the center drill from the lathe chuck, and drill the holes in the steady accordingly, using the same registration as before.

Now unbolt the steady from the cross slide, and, in the vise, press the inserts into the desired holes, from the Q side (Fig. 89). Use a flat steel block to keep the insert lined up squarely and protect it from the vise jaws.

Now mount the arbor in the lathe as shown in Fig. 90. Fit the chamferer in the tailstock

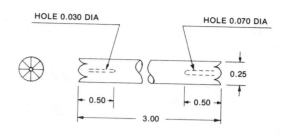

Fig. 87. Chamferer

chuck, and rest the free end of the arbor in the end of it. At about 800 RPM, press the chamferer lightly against the arbor with the tailstock screw until a small chamfer appears. Although the hole in the chamferer provides pivot clearance, since no pivot is present, it does not matter which end is used. Do not make a large chamfer, or the end bearing will be too small.

Bolt the steady onto the cross slide and line up the 0.040" hole, using a dead center in the tailstock. Set the cross-slide micrometer dial to 0. Slide the carriage along until the arbor fits snugly into the insert (no. 1), and lock the carriage. Note that the chamferer will have cut a taper of some 79° on the arbor, so that only the outside shoulder of the arbor contacts the 60° taper of the insert; this prevents any burring of the end bearing.

Put a small drop of oil in the insert, and then use a no. 60 drill (0.040") in the tailstock to make a centering hole in the end of the

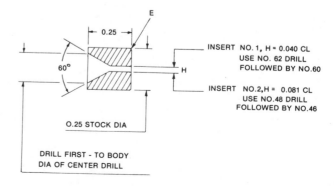

Fig. 88. Repivoting Insert

arbor, as shown in Fig. 90. Transfer insert no. 2 (0.081″) to the arbor, so that the hole for the new pivot is ready for drilling.

Before drilling holes such as this, there are certain aspects to consider. Holes drilled in the usual manner are slightly trumpet mouthed; this is because the dwell time of a drill is greatest at the entry end of a hole. In an arbor, such a hole would result in a tiny annular gap around the inserted pivot, which would retain dirt and allow the pivot to flex somewhat.

To obviate this problem, *two* drills are used. The hole is first drilled a few thousandths of an inch undersize, so that the second and final drill has very little metal to remove and can therefore cut very quickly, with a short dwell time.

In drilling for the new pivot then, the original pivot diameter will be assumed to be, for example, 0.048″. For a reason given later, it is a good idea to make the new pivot slightly larger—say, 0.053″ in diameter. A press fit for a 0.053″ pivot requires a finished hole of close to 0.052″; this is best drilled in two stages, for the reasons given above (dwell time, and so on).

Mount a no. 56 drill (0.0465″) in the tailstock chuck, and drill the initial hole, about 0.37″ deep, in the arbor (with the drill passing through insert no. 2). Change the drill to a no. 55 (0.052″) and repeat the process—rather more rapidly this time. Do not drill the hole any deeper, and remove the drill as soon as it bottoms in the hole. This will give a hole only a few tenths of a thousandth of an inch over 0.052″, and so provides a good press fit for the new pivot. While these figures are typical, a larger arbor would, of course, require a larger pivot and a deeper hole.

Now slide the tailstock away and remove the steady. Check for burrs and apply a polishing stick to the end of the arbor, as required, to obtain a smooth finish. (This can also be done as shown in Fig. 86.)

File a small chamfer, for a lead-in, on the end of a short piece of 0.053″-diameter pivot wire, and tap it gently into the arbor. A vise-held split-plate is an alternative to the chuck for holding the arbor. Note that it is unwise to involve the remote end-bearing, as damage would be the likely result.

Cut the pivot almost to length, and with the

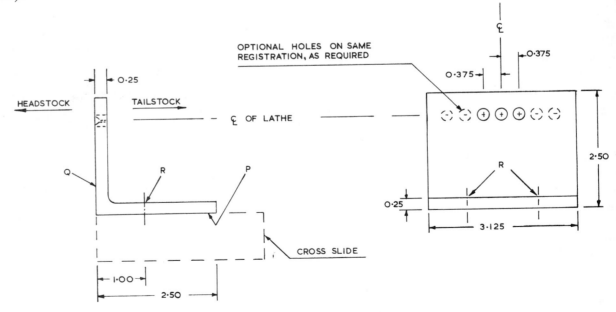

Fig. 89. Repivoting Steady

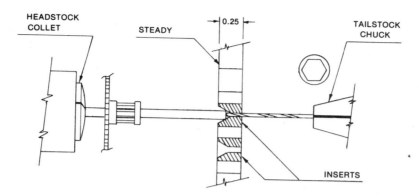

Fig. 90. Repivoting an Arbor

arbor mounted as shown in Fig. 85a remove any small burrs by coning the end of the new pivot slightly with a fine file, as shown. In some cases, the closed runner of Fig. 85b may be needed for this. Give both pivots a final polish.

The reason for making the new pivot larger than the original is that this sometimes obviates having to bush the pivot hole in the plate. The pivot hole will naturally be somewhat worn and so require attention. Broach the hole lightly to suit the new pivot (as given in **Bushing**). If, after broaching, the hole is still out of round—even slightly—then it must be bushed **(Bushing: Plates)**.

Note that insert no. 1 (0.040″ diameter) with its no. 60 drill is used only to make an accurate centering hole in the end of the arbor. Transfering insert no. 2 (0.081″ diameter) to the arbor permits the drilling of centered holes ranging in diameter from 0.040″ to 0.081″. Insert no. 2 therefore may be called a clearance insert, as it provides clearance for many drill sizes. Other pairs of inserts can be made for holes above or below this range.

Conventionally, these repivoting holes are made by drilling a one-shot custom hole in the steady itself. The steady is sometimes known as a flag; no insert is used. With this method clearances are small, and swarf buildup, unless continually cleared, is liable to break drills. I consider the use of one or two permanent hard inserts, on a repeatable registration, to be superior to the common way of drilling and countersinking a new hole for each new pivot.

Tailstock runners can be made of wood (birch), but brass lasts longer. Take a few inches of brass rod ¼″ or ⁵⁄₁₆″ in diameter, and face off the ends in the lathe. Drill a central hole about ¼″ deep with a no. 60 or other small drill, and then file half the rod away for about 0.8″.

In certain awkward configurations that are difficult to mount, or when deep rusting appears, it is often best to make a cap pivot (see Fig. 91). Often, when an old clock is rediscovered—say, after being in a barn for a few decades—its movement is severely rusted. When this rust has spread to the pivots, it often takes the form of tiny axial cracks which resemble fine, black lines; these cracks sometimes extend into the end bearing and arbor, and will not polish out, though they may appear to do so. In such a situation, a cap pivot is a good alternative to a new arbor.

Shorten the arbor by about 0.06″, and clean

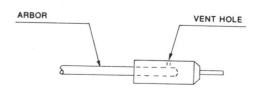

Fig. 91. Cap Pivot

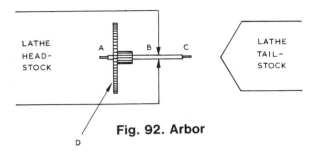

Fig. 92. Arbor

it well. Turn up a suitable cap with a slightly oversize pivot, and make a vent hole with a no. 60 or other small drill (Fig. 91). Chamfer the cap to give a reasonable end bearing; if it were left full size, oil drag would be excessive. The cap pivot must be only finger tight on the arbor; it is not a force fit. Try out the arbor and cap pivot between the clock plates; when you have achieved the correct length, flux the joint and soft solder it in position. The vent will prevent air locks and bubbling.

Mild steel is adequate for most pivot caps, but a hardened version is feasible when made from drill rod.

Certain arbors, such as those of lifters and anchors, are very difficult to chuck without bending or removing a hook or crutch. However, because they only turn partially and do not usually exhibit much wear, they can be polished by hand.

Plewes chuck: I designed this chuck to overcome a very old and awkward horological problem—that is, how to mount a typical clock arbor, as shown in Fig. 92, so that the pivot C can be machined, replaced, or otherwise worked upon. If the arbor in Fig. 92 is reversed end-to-end, it is usually easy to grasp it at point B in a lathe chuck and then machine or polish the pivot at A. The problem of dealing with pivot C, however, does not seem to have ever been solved in a really sound mechanical way. To polish or burnish pivot C, a Jacot tool is generally used, but when the A end cannot be chucked firmly—and in most cases it cannot—this method is unsound, since damage can easily result. If pivot C is to be replaced, then it is good practice to use a fence with a suitable hole in it, but the difficulty of chucking A remains. Driving between centers

Fig. 93. Plewes Chuck

Fig. 94. Chuck Mounted on Lathe

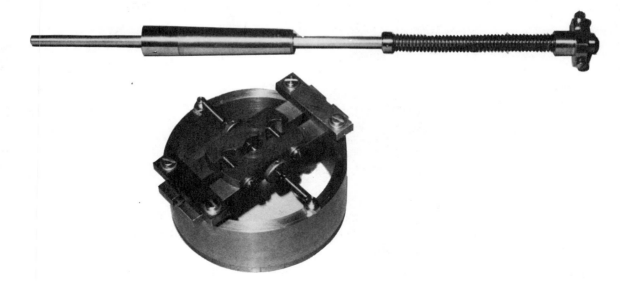

Fig. 95. Chuck Center Shaft

does not help, because then neither pivot is accessible.

If pivot C is to be machined, and so on, with the same ease as pivot A, it becomes necessary to "reach over" the gear D in some way to grasp the arbor at B, so as to obtain sufficient rigidity. This basic concept must then meet the following parameters:

1. The arbor must rest on the centerline of the lathe.
2. The chuck must be able to accept the tapered, out-of-round, irregular arbors found in old longcase clocks.
3. The entire chuck mechanism must not be more than ½" thick.
4. The chuck must be in rotational balance.
5. It must be feasible to make the chuck on an average bench lathe, using commonly available materials and reasonable care and tolerances.
6. The chuck must be easy to set up and maintain.

A conventional scroll-type self-centering mechanism would be difficult to design to meet requirements 2 and 3. In view of this, I decided to mount the arbor between centers, grasp it without disturbing it, and then slide the tailstock away, thus leaving the pivot free for machining (Fig. 94).

The new chuck itself (Fig. 93) mounts onto the lathe's conventional 4", three-jaw, self-centering chuck, as shown in Fig. 94. Fig. 95 shows the chuck and a spring-loaded, female headstock center shaft with front and rear bearings designed to fit a Myford ML7 lathe. When the chuck is to be fitted to a different lathe, these two bearings must be altered to suit. A good-quality brass-bushed radio knob can be used to retract the shaft, as required, to facilitate setup. A female tailstock center is also required; this can be 2" of ³⁄₁₆"-diameter drill rod with centers of different sizes at each end. Both centers should be hardened and tempered to a light straw.

The following drawings give dimensional details of the various components. The accommodating action of the hemispherical jaws obviates any need for unduly tight tolerances; nevertheless, considerable care must be used throughout.

Fig. 96 is an assembly drawing, showing

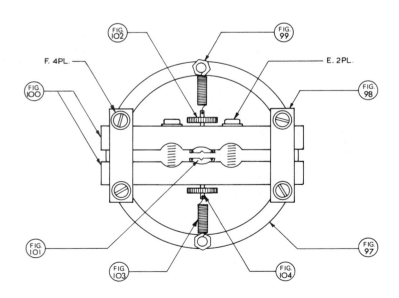

Fig. 96. Chuck Assembly

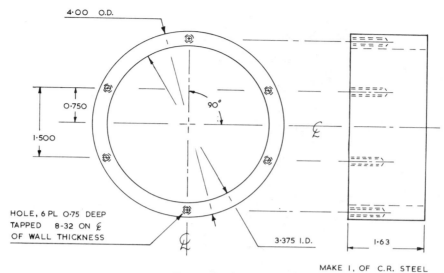

Fig. 97. Cylinder

how the chuck is assembled from the parts shown in Figs. 97–104. A 4″ cylinder with an I.D. of 3.375″ will accommodate the largest clock gears commonly found; Fig. 97 gives its dimensional details.

Fig. 99 shows two studs which are screwed in by hand until 0.32″ projects from the cylinder end. The nuts are then tightened.

Fig. 100 depicts clamp bars. These are made from 0.5″-square stock. All bumps and burrs must be removed and all surfaces must be flat and true. First, accurately align the bars and mark them out. Then clamp them 0.25″ apart by means of a 0.25″ × 0.50″ steel spacer, to enable the recesses G to be cut. Use a 0.50″ drill on the centerline of the spacer. Then drill the holes H and J. It is best to do this separately. Accuracy is important here. Some runout may occur if the bars are clamped together for drilling; however, the upper block can be used as a guide for tapping the lower one.

Now mount the blocks together, center them in a four-jaw chuck, and machine the ends down to 0.250″. Parallelism and uniform

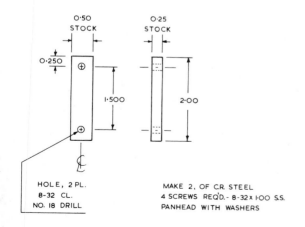

Fig. 98. Straps

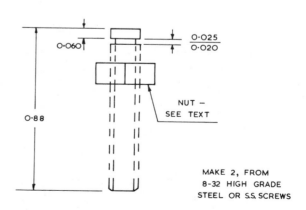

Fig. 99. Studs

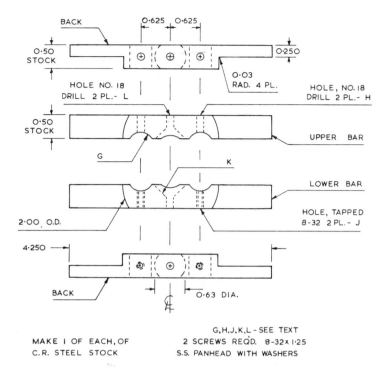

Fig. 100. Clamp Bars

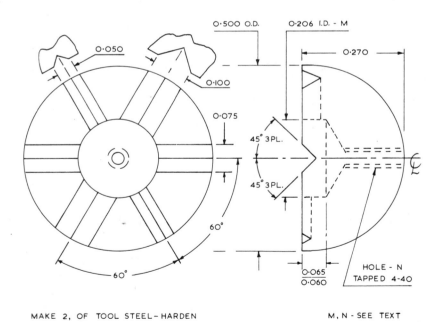

Fig. 101. Hemispherical Jaws

thickness are more important than an exact 0.250″, so painstaking setup is essential.

Set up a clamp bar accurately in a four-jaw chuck, and machine out the central, conical jaw recess K and central hole L. Assuming that the back surface of the first bar to be machined abuts jaws no. 3 and no. 4 of the four-jaw lathe chuck, then loosen only jaws no. 1 and no. 2 for removal. This leaves jaws no. 3 and no. 4 as a reference for the back of the second bar, thus facilitating setup and ensuring that the recesses are equidistant from the backs of both clamp bars. The dimensions given allow the hemispherical jaws to swivel a little over 6° from the centerline.

Fig. 101 shows the chuck's hemispherical jaws. Mount a few inches of 0.50″ drill rod (tool steel) in the lathe, face both ends smoothly and drill holes M and N. The three grooves are best made using a dividing head, a fly cutter, and a milling arbor. However, there are other feasible ways of cutting slots on the lathe. One idea is to use a fly cutter in a bar, turned by the chuck and steadied by the tailstock. A small jig of 0.50″ plate would then hold the stubs of the jaws on the cross slide. Protractor-drawn lines, angled 60° from the centerline of the cross slide, would serve to orient the stub reasonably accurately. The shallowest grooves (0.050″-wide), would be cut first and then rotated to the 60° mark, and so on. Other methods may suggest themselves, but any method that does not involve a dividing head and milling arbor admittedly requires ingenuity and more time and work in setup.

After cutting the grooves, cut the stubs down close to 0.28″ in length, and make the mandrel of Fig. 110 (p. 208). Carefully grind a form tool to conform to a quarter-section of a 0.50″-round rod, using a 0.31″ or 0.37″ high-speed lathe tool. The tool must be undercut, with its edge stoned smooth. No light should show when the finished tool is held against a smooth 0.50″-diameter metal rod.

Chuck the mandrel and screw on one of the

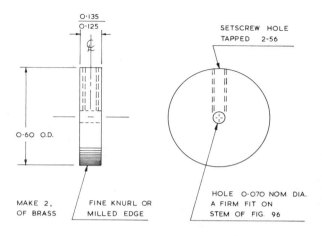

Fig. 102. Knurled Discs

stubs. Using a sharp tool and a light cut, gently turn down the stub to its correct height and approximate shape. Reduce the lathe speed to about 35 RPM, and with a light cut use the form tool to finish up. Rub the resultant hemispherical jaw to a smooth finish with emery paper. Repeat the process for the second jaw.

Heat the two jaws—no more than red hot—and quench them. Clean and polish them again on the mandrel, and then temper them carefully to a light straw.

Fig. 104 depicts the stems. Machine the underside of the screw head to a snug fit with the angle at the bottom of hole M (Fig. 101), making sure that the screw head lies below the bottom of the deepest groove. Machine the threads down, and then tighten the screw firmly into the jaw. The flat P (Fig. 104) is now

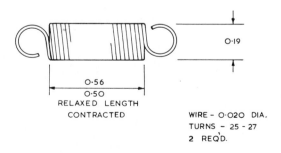

Fig. 103. Chuck Spring

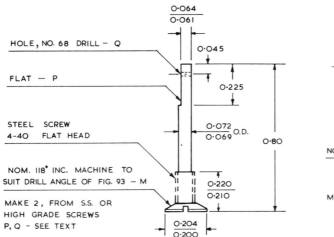

Fig. 104. Stems

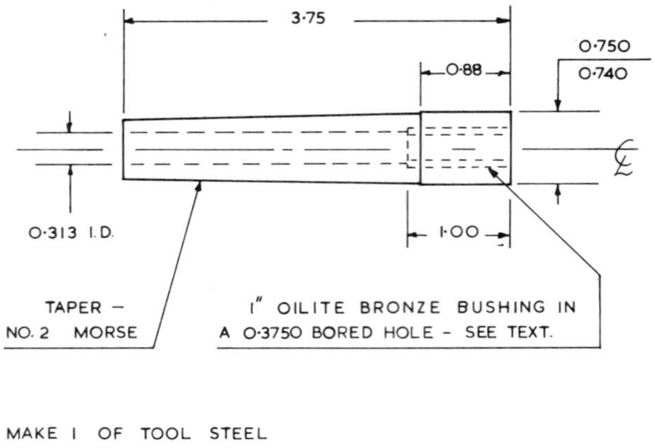

Fig. 105. Center-Shaft Front Bearing

filed on, directly opposite the end of the 0.075″-wide groove. Drill the small hole Q.

Fig. 105 shows the center-shaft front bearing, fully machined. Instead of making it as shown, it is possible to modify a mild-steel dead center (No. 2 Morse taper) to achieve the same configuration. (A brass insert must then be used to obtain the dimensions given.) In either case, the central hole for the bronze bushing must be bored, not simply drilled. This is necessary to achieve the required accuracy.

The center shaft in Fig. 106 must be a straight piece of precision shafting, with the female center hole machined in it as concentrically as possible.

Fig. 109 shows the center-shaft rear bearing, which must initially be a firm fit in the rear end of the headstock. Do the machining and drilling and tapping first, leaving hole R a firm no-play fit on the center shaft. Cut slit S last.

The chuck is assembled as shown in Fig. 96. All six screws have washers under their heads. The set screws of the milled discs grip the flats on the stems. The springs of Fig. 103 should be oriented so that the set screws are normally facing centrally outward. They can then be

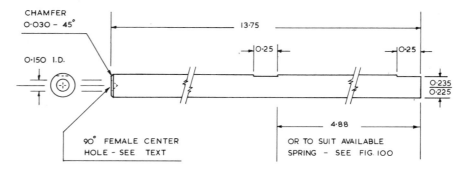

Fig. 106. Center Shaft

Plewes Chuck

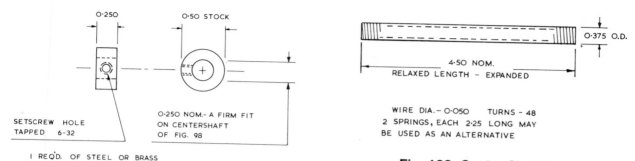

Fig. 107. Collet

Fig. 108. Center-Shaft Spring

used to indicate matched groove sizes, by turning them both together as required. The springs hold the jaws apart while retaining them in their recesses.

Setting up the device only takes a few minutes. Make sure that there is no swarf or dirt inside the entire lathe headstock, then clean the center-shaft front bearing and insert it firmly into the lathe headstock, leaving the lathe's three-jaw chuck in place. Assemble the center-shaft rear bearing and springs as shown in Fig. 95. Now feed the shaft into the rear of the lathe headstock spindle, and carefully insert it into the rear of the front center-shaft bearing. Push the center-shaft rear bearing home fully, keeping the T-screw of Fig. 109 on top. The female center will now appear in the middle of the three jaws of the lathe chuck. Now turn the U-screw *gently*, and the rear bearing will expand to grip the headstock spindle; simultaneously, its grip on the center-shaft will relax a little.

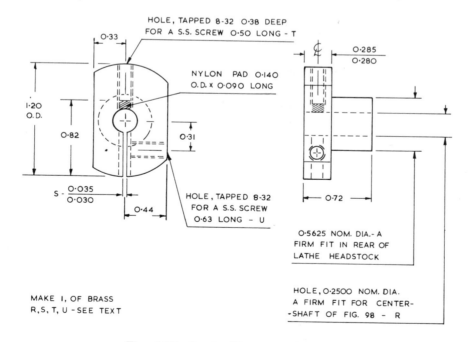

Fig. 109. Center-Shaft Rear Bearing

Finally, install a suitable small, fixed female center in the tailstock.

Now take the chuck assembly and slide the desired arbor between the hemispherical jaws, removing one of the clamp-bar screws if necessary. Select a suitable pair of grooves in the hemispherical jaws by rotating the brass discs (using the set screws as indicators), and then tighten the two clamp-bar screws E (Fig. 96), until the jaws embrace the arbor loosely.

Now fit the cylinder on the lathe chuck, as in Fig. 94. Mount the arbor between the centers, and slide it along against the center-shaft spring as required, until the damaged pivot is close to the hemispherical jaws, using the tailstock screw. When the correct position has been found, the center shaft can be locked, if desired, by means of the T-screw in the center-shaft rear bearing; this is not essential.

Now carefully adjust the four strap screws F (Fig. 96), until the straps rest lightly and evenly on the clamp bars. It is essential to do this carefully; the bars should be just free enough to move under the straps, but without any free play. The strap must not grip one bar more than the other; they must be level. Keeping the clamp bars sensibly parallel, move them slightly, as necessary, and then screw them together until the jaws grip the arbor firmly. Check the freedom of the bars under the straps, and then tighten their four screws. The force applied by these six screws is entirely sufficient to hold a clock arbor rigidly in position for machining. Do *not*—repeat *not*—tighten them with excessive force; there is absolutely no need. Excessive tightening of these screws will only produce damage and unwanted wear.

Socket-head screws may be used instead of slot heads, but they protrude farther than pan heads. In addition, swarf and dirt tend to plug them.

Aside from setup, the practical accuracy of alignment also depends upon the shape and condition of the arbor pivots. Original pivots are by no means always concentric with their arbors, and this should be corrected along the way. It is sometimes best to remove a bad pivot, and use the end of the arbor, faced-off true, as a center reference. The accuracy of the tailstock female center is also a factor; this should be of hardened tool steel, so that it cannot easily be deformed. In practice, the total accuracy obtained is excellent—at least equal to that of a good self-centering, three-jaw scroll chuck.

In any case, even small run-outs of some 0.002" can easily be corrected by slackening the strap screws and either tapping the clamp bars lightly with a screwdriver handle or manipulating them by hand.

Longcase seconds arbors are dealt with as given above. The rear pivot of longcase minute arbors can be machined in the usual three-jaw

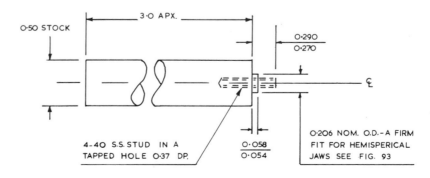

Fig. 110. Mandrel

chuck by means of a thick-walled brass collet. This is drilled to a firm fit on the pinion and then slit to allow contraction; the slit should lie between two pinion leaves. The front pivot of this arbor (usually very awkward to deal with) is easily machined in the Plewes chuck. Slide the tapered minute-hand end of the arbor between the hemispherical jaws, and mount the arbor between the female centers of the center shaft and tailstock, with the pinion end in the tailstock. Set the front pivot in a machinable position, close to the hemispherical jaws, and clamp them on the arbor, using moderate pressure. Run the lathe and correct any run-out as given above. Tighten the clamp bars and strap screws, leaving the tailstock center in place. Now machine the front pivot true, from left to right, and polish it as usual.

Since it only requires seven minutes or less to mount these otherwise most troublesome arbors in the lathe for machining, this new chuck will well repay the time spent on its construction. I find it most useful.

Reduction gears: These are usually known as "motion wheels" or "motion work." They reduce the once-an-hour speed of the minute hand by twelve-to-one to the once-every-twelve-hours speed of the hour hand. A further reduction of two-to-one to once a day can be used to drive a thirty-one-toothed gear to operate a day-of-the-month calendar arbor.

Remontoire: This is a "remounter," used to rewind a small weight or spring periodically to keep a clock running. Remontoires are often contact-operated electrical solenoids, which raise a fallen weight, as necessary, to keep a clock running. Thus a constant-force drive is maintained by the use of somewhat less-constant electric power.

Soldering: This is so widely practiced in one form or another that its importance can hardly be overemphasized. It seems to cause more people more trouble than any other technique. Many even claim soldering ability to be an innate gift—presumably from the High Temperature Department—but I trust that this is not so.

There are two basic principles involved with all forms of soldering, and until these are thoroughly understood, accepted, and implemented, progress is almost impossible. The two principles are scrupulous cleanliness and the application of sufficient heat. There are no short cuts.

Soldering is a metal-to-metal proposition in the most literal sense; solder will not bond to dirt or metal oxide. Flux is quite widely thought to be a dirt remover, whereas it is really an air excluder whose purpose is to prevent oxidation of clean metal as heating proceeds, so that solder can effect a bond. Flux, even acid, is no substitute whatever for a thorough cleanup.

Because the transfer of heat takes time, there is a limit to the speed at which soldering can be accomplished. It is most unwise to try to rush matters.

Soldering: *Silver soldering*. This is the joining of ferrous and nonferrous metals with a solder made up of silver and copper plus a little zinc, etc., depending upon application. The flux used is borax, and red heat is required; because such heat readily anneals (and thus softens) copper alloys such as brass, silver soldering is most useful in bonding steel to steel. Silver soldering and brazing are two of the techniques known as "hard soldering."

As an example, on a longcase movement two areas which often require silver soldering are the steel hands and the pendulum crutch, which often suffer transit damage. Soft soldering is inadequate in both of these instances.

To repair a broken pendulum crutch, file both broken sections bright and clean, and in the rectangular fork file a slot which is a firm fit for the end of the arm: This will help hold it as it is soldered. Flux both parts with borax and fit them together. The end of the arbor can be mounted in a vise, so that the fork rests on small, hard board on the bench, for no movement of the joint must occur until it is well set.

An ordinary propane torch is adequate for such small jobs, but do not hold it too close to the job; use the hottest part of the flame—the tip of the blue cone. When the parts are red hot, apply a small amount of silver solder. It should run immediately into the joint. Use the solder sparingly, for a buildup is undesirable, both structurally and financially. When the job has cooled, hold it under boiling water to remove flux residue, a hard glaze which will speedily ruin a file. Dry the joint, and clean it up with a wire brush and emery cloth; do not leave any residue around the area, or corrosion may result.

Soldering: *Soft soldering.* This is the joining of ferrous and nonferrous metals with a solder made of various proportions of tin and lead. The flux used is resin, and temperatures of 370°–450°F. are required for the usual 60:40 tin-lead solder. Either a flame or soldering iron can be used to apply this heat; the choice is a matter of experience.

Soft soldering is attempted more frequently than hard soldering, so most of the frustrations experienced seem to involve the soldering iron. Those who fail to graduate often achieve a master's degree in profanity instead.

A big advantage of soft soldering is that it is done at a temperature much below the 700°–800°F. which anneals copper and its alloys, and thus preserves their work hardening. The strength of the joint is much less than that of hard soldering, but is adequate for a wide variety of jobs. Among the metals which will accept soft solder are copper and its alloys such as brasses, bronzes, gun metal, phosphor bronze, beryllium bronze, beryllium copper; and steel, nickel, tin, lead, zinc, gold, silver, and other noble metals. Certain stainless steels will also soft solder, but acid flux may be required; there is no need of acid flux on any of the nonferrous metals, and its use should be avoided wherever possible, due to corrosive residues.

Gold and silver watchcases should never be soft soldered, because this detracts from their value and makes proper repair using gold or silver more difficult.

Copper and its alloys solder easily, provided they are first thoroughly cleaned with emery cloth, a file, or wire brush, or even turned on a lathe. Zinc must be very thoroughly scraped down to the bright metal and fluxed before it will solder; if any pits are left in it, the solder film will be broken. Whatever the metal, do not polish the soldering surface; it must be very clean, but the fine scoring of the emery cloth or file helps to form a bond, and therefore should remain. Flux-cored solder is generally adequate, but paste flux is useful in many cases. Acetone readily removes flux residues, but contact with it and inhalation should be avoided.

Cleanliness also applies to the soldering iron, so having cleaned and fluxed the job, clean the iron and tin it with solder and flux. If it is a bare copper iron, file out the pits and remove any black scale; a dirty iron will not heat the job properly. The solder itself must also be clean.

The contact area between the job and soldering iron is quite small, and little heat can transfer. A small puddle of molten solder in the contact area, however, will conform to the shape of both iron and job, and a useful heat transfer will then take place. What is often overlooked is that this transfer takes time; the iron must be left on the job, in its solder puddle, until added solder flows throughout. Still keeping the iron in place, position the parts where you want them and hold them there with a scriber, screwdriver, plier tips, or other device, and only then remove the iron. Keep the hold-down tool in place, holding it dead still until well after the solder has appeared to set; this may be up to fifteen seconds, depending upon the size of the job. The visible solidification is only a surface effect; time is again necessary for the job to cool off and so allow the joint to solidify throughout.

Most flux residues will promote corrosion sooner or later, so it is best to remove them thoroughly. This particularly applies to acid fluxes and acid-core solders; however, these are seldom required in clock repair.

The foregoing information is given to promote sound soldering, and to thin the ranks of those who have erroneously arrived at the conclusion that they cannot solder.

Spandrels: the decorative castings in the corners of a brass dial. Painted scenes in a white dial are known as "corner decorations" or "corner scenes."

Split stake: A split stake, or split hole plate, is very useful in dealing with such parts as collets on arbors (see Fig. 4). It basically consists of two strips of metal, dowelled together side by side, with a series of holes of various sizes drilled along the line of contact. The strips can be separated and then closed up again with an arbor grasped in one of the holes, so that a collet or shroud may be moved or riveted, and so on. A typical split stake may be made of brass or mild-steel strips, ¼" × ½" × 6", as shown in Fig. 111.

Two steel dowels, made from old drill stubs about 0.10" in diameter, are fixed into section A of Fig. 111 and slide out of section B to enable separation. When making the stake, line A and B up carefully side by side and fit the dowels, before marking out the split holes. These are positioned to the nearest five thousandths of an inch on the line of contact, and are center punched and drilled in the usual way, with A and B clamped firmly together in a drill vise. The hole sizes given range from 0.040" to 0.246", but may be changed as you prefer.

After the surfaces are brightened with fine emery cloth, a random machine finish looks well. This is done in the drill press with a small strip of emery cloth wired across the end of a piece of ⅜"-diameter birch dowel, using a fairly high speed.

Suspension spring: the flat spring which supports the pendulum and allows it to swing. (See under **Pendulum**.)

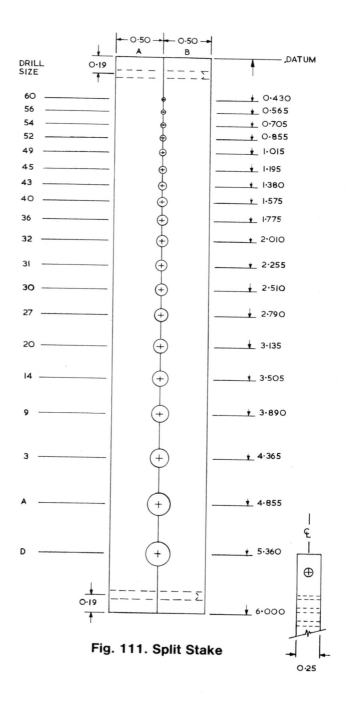

Fig. 111. Split Stake

Timepiece: a time-only, one-train clock.

Tooth replacement: *Gears.* Gear teeth usually suffer damage when a mainspring breaks, which makes a good method of tooth replacement essential. If several teeth are involved, file out a slot, and soft solder into it a piece of brass slightly thicker than the gear, as shown in Fig. 112a. If, for example, five teeth are broken, clamp a scrap of tinned steel from a can alongside nine good teeth, and scribe their outlines on the steel, thus making a template.

Remove the template and carefully file out the tips of two teeth, at each end of it. Using these as references, line them up with two good teeth at each end of the damaged section (as shown between P and Q in Fig. 112a), and clamp the template in position. Now file out new teeth to the scribed outline on the template.

If only one or two teeth are broken, file the remnants flat and cut separate slots for each tooth, as shown in Fig. 112b. Keep the sides of the slot parallel, and file out a thick brass insert which can be squeezed into the slot with pliers or tapped home if a spoke interferes (Fig. 112c).

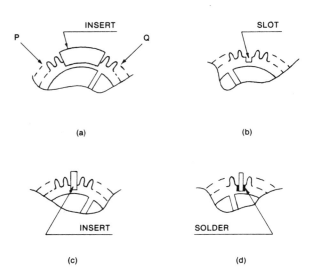

Fig. 112. Brass Gear-Tooth Replacement

The fit is important, for silver soldering cannot be used here, because it would anneal (soften) the brass and ruin the gear; therefore soft soldering must be used. Because soft solder is not structurally strong, a brass-to-brass contact is essential for the new tooth. A thick fillet of solder, as at Fig. 112d, would allow the tooth to loosen in time; the solder at one side would become compressed, and a gap would form at the other.

When soldering the insert in place, hold the arbor in the vise, with the gear horizontal. Hold the iron up to the underside of the gear, and when the fluxed joint warms up, apply a little resin-cored solder to the topside of the joint. This will cause the solder to flow through the joint to the iron. Only a small amount of solder is necessary; remember that all excess solder has to be filed off.

When the joint is cold, remove all flux residue and file the new tooth, or teeth, to shape. File down both sides flush, so that the only solder visible is a thin outline of the insert. This repair is as strong as other teeth; when polished with fine emery cloth, it is not readily seen at a glance.

See that the repaired gear will drive the next pinion, without any sign of jamming, by mounting the two arbors between the clock plates and spinning them together. Check that there is enough backlash when each trundle engages the new tooth, and file it a little if necessary.

Tooth Replacement: *Going Barrel.* When a mainspring breaks inside a going barrel, it generally breaks two of the barrel teeth and damages others. A slightly bent tooth can often be straightened, but the broken ones, of course, must be replaced.

Some people replace a tooth with a steel screw, which they file flat to form a rough sort of tooth, but this is a poor repair, for the "tooth" is narrow and very hard on the next pinion.

A much better way is to cut slots and use an insert, as in a gear, but because of the barrel they must be angled, as shown in Fig. 113b. Such slots break through a little into the inside of the barrel, but the inserts do not touch the mainspring, because it is held away from the barrel end by the arbor bushing.

There is no need of a template for two teeth, but, as with a gear, the inserts must be a firm fit in their slots and thick enough to be filed down to form teeth.

Many barrels are soft soldered to their gears, so it is best to disturb the joint as little as possible. Mount the barrel halfway down in the vise, with the inserts uppermost, using two shims of wood to prevent heat loss. Flux the inserts, and apply a torch flame to them inside the barrel. As soon as heating allows, touch a little resin-cored solder to the job, and when it runs through withdraw the flame. (Note that a torch is used because it is faster; an iron would be more likely to melt the whole barrel joint.)

When the barrel is cool, remove all flux residues, file off surplus brass, and file the teeth to size. Now assemble the barrel and its arbor, together with the next arbor in the train, between the clock plates, checking that the new teeth mesh correctly with the pinion and have adequate backlash.

When carried out well, this repair is strong and sound; as with the gear repair, there should be little evidence of its presence.

Tooth Replacement: *Wooden Gear.* The gearing of a wooden works clock is naturally not as strong as its brass counterpart. This is not always appreciated, and breakage here is somewhat more common, often occurring during winding.

The conventional method of repair involves cutting a slot or drilling a hole and gluing a new tooth in place. However, as epoxy resin is now available, it makes sense to use it instead. It can bond new teeth in place, or be moulded to form the teeth themselves, without making them separately of wood.

To make a mould for forming new teeth, take a small amount of flexible, rubbery modelling clay—available at toyshops—and knead

Fig. 113. Going-Barrel Tooth Replacement

it firmly around a suitable number of undamaged teeth (see Fig. 114). When the clay is shaped, cut one side of it away, as shown in Fig. 114a. Remove the mould from the sound teeth, and make sure that the repair is free from oil, wax and grease, or the resin will not bond to it. Cut out a slot in the gear as a key for the new tooth, and set the mould in place (Fig. 114b). Make sure that the mould contacts the gear closely; if necessary, clamp it lightly to prevent leaks.

In order to stain the new tooth to resemble the wood of the gear, load the epoxy with fine sawdust—it must be fine—with no small shavings in it. Five-minute epoxy resin involves some fast mixing, so it is probably best to use resin with a longer setting time, in order to achieve a homogeneous mix and a good bond. A heat lamp will usually speed up the setting process, if desired.

The exact amount of fine sawdust loading is

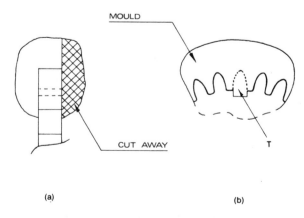

Fig. 114. Wooden Gear Tooth Replacement

hard to specify; it is best to experiment a little. It does not appear to weaken the resin; there is no problem as far as strength is concerned. I have used such mixes for about twenty years, and no deterioration has been evident. A 1964 example in my possession still holds up.

When the mix is satisfactory, fill the tooth area T (Fig. 114b), using a toothpick or wooden sliver, the fine point of which can extract any air bubbles. Add to the mix, if necessary, so that its surface is slightly above that of the gear; this will take care of any shrinkage which may occur during setting. Because many resins take a day or two to cure, it is best to leave the mould in place for at least a day.

Take the mould off carefully—the original teeth may well be weaker than the new one—and file off any surplus resin so that both sides of the gear are flat. Smooth the new tooth with fine sandpaper. Test it in the movement, and trim it a little if required. Stain the tooth to suit.

Wooden pinions can also be repaired in this manner.

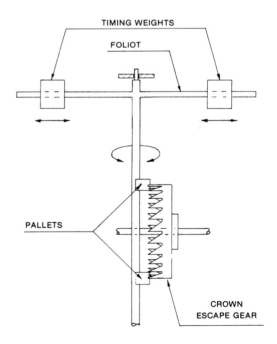

Fig. 115. Verge Escapement

Trains: Also called geartrains, these are assemblies of gears, as in a clock.

Trundles: the pins in a lantern pinion (see **Pinions: Lantern**).

Unlocking Hook: a lifter in a strike train (see Figs. 7 and 17).

Verge: an anchor, or pallet assembly.

Verge escapement: This type of escapement dates back about six hundred years. It is not often found today, for it was superseded by the superior anchor type, about three hundred years ago, so it is now chiefly of historical interest. Fig. 115 shows the general idea of this old escapement: The pallets are actuated by the teeth of the crown-like escape gear, and as a result the foliot, with its two adjustable weights, swings back and forth to act as a governor.

In a later version, the escapement was turned 90°, and a contrate gear was used to drive the now vertical crown-gear arbor. One of the weights was removed from the foliot, thus forming the short bob pendulum, another rarity today.

Verge-making: Kits are available from which new verges can be made. They consist of a strip of tool steel, a small piece of brass for a saddle, and a wire for the crutch. It is best to copy the dimensions of the original verge, if possible, and bend the strip to suit. Drill two holes and mount the saddle and crutch, adjusting the verge as given above under **Escapements** before hardening it. Strip-pallet verges such as these are easier to make and adjust than are the solid anchors of longcase clocks.

Warning: This is the start of the striking sequence, when the unlocking pin has raised the unlocking arm so that the strike train is ready to run at the hour. The warning usually occurs during the last ten minutes of the hour. (See Figs. 7 and 17.)

Weights Clock weights come in many shapes and sizes. A large proportion are made of various types of cast iron, some much too hard for high-speed steel drills to penetrate. The weights of longcase and ogee clocks are typical examples. The hooks or wire loops which are cast into them are often off center, and do not allow the weights to hang straight; this may cause needless damage to surrounding wood. For example, a longcase weight may hit the bottom edge of the case-door opening.

Before installing such a weight, hang it on a loop of coathanger wire, and then bend the hook as required until the attachment point is on the centerline and the weight hangs vertically. This is easily done by clamping the hook in the vise jaws and gently moving the weight itself by hand. The loop of an ogee weight can be altered, lengthwise or sideways, by a few light taps with a hammer. Weights with broken hooks and that resist drilling need not be thrown away; new hooks can be arc-welded onto them.

A few weights, such as those for an American 30-hour four-column clock, can be made by riveting together short lengths of square-section steel, but in most cases weight-making involves the use of molten lead.

Replacement brass-shelled weights are sometimes needed for Viennas and American regulators. It is essential to know the poundage required to drive the clock; from this the dimensions of the shell can be calculated. The volume of a cylinder is given by

$$v = r^2 L$$

and the volume of the shell is given by

$$v_s = (r_o^2 - r_1^2) L$$

where

v = volume of cylinder in cu. in.
r = radius of cylinder in in.
π = 3.14
L = length of cylinder in in.
v_s = volume of shell in cu. in.
r_o = outer radius of shell in in.
r_1 = inner radius of shell in in.

Note that

1 cu. in. of brass weighs 0.3 lbs.
1 cu. in. of lead weighs 0.41 lbs.
1 cu. in. of steel weighs 0.28 lbs.

Fig. 116 shows a good method of making up a weight.

Turn up two 0.062"-thick brass discs, of the

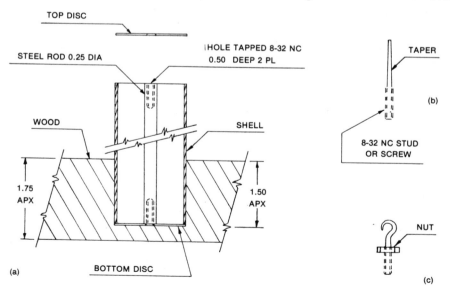

Fig. 116. Brass-Shelled Lead Weight

same diameter as the shell, and drill a central hole in them with a no. 18 drill (8-32 clearance). Countersink one of them to form the base of the weight. Cut and face a piece of 1/4" steel rod to the same length as the shell, and then drill and tap both ends of it 8-32 NC × 1/2" deep.

With an expansive bit, drill a press-fit hole for the shell in a piece of scrap lumber, and assemble the components as shown in Fig. 116a. Melt and pour the lead, keeping the steel rod centered at the top of the shell; the weight will look untidy if the top disc is off-center. Allow the weight to cool and then take it out and remove any flashings. Hammer or file both ends of the lead as necessary, so that the discs fit snugly. The heat may have tarnished the brass shell, but metal polish will soon restore it.

Take about 1½" of 8-32 NC brass studding, or a suitable screw, and turn a taper on it, leaving about 5/8" of thread (Fig. 116b). (A steel hook would be stronger, but does not look quite in keeping with the brass shell.) Bend the tapered section into a well-shaped hook, and add a brass nut (Fig. 116c). Fit the hook through the top disc and screw it into the rod as far as it will go; tighten the nut to provide locking. Polish and lacquer the entire weight.

Where good appearance is not a factor, wooden moulds are quite suitable.

Instructions for making a weight for a banjo clock are given on page 36.

Wheels: As used in clock circles, *wheel* is a widely accepted and totally incorrect term for a gear of a clock movement.

Workshop safety: Although I cannot mention all the mistakes that cause damage to people and machinery, it seems desirable to mention a few that sometimes do *not* come to mind—until it is too late.

Asbestos. The small pieces of hard asbestos sheet sometimes necessary in silver soldering should be handled with care, and not stored where abrasion is likely. Most asbestos fibers evidently come from asbestos wool, but wherever they come from, they constitute a serious health hazard.

The chisel is one of the tools that we learn to use, often incorrectly, rather early in life. It must always be kept very sharp, and therefore it can inflict a deep cut should a slip occur. Always try to carve wood away from your passive hand, so that if the chisel skids it is out of the way. Even a screwdriver is dangerous in the same situation.

Chuck keys. Never leave a key in the chuck of a machine, even for an instant. Look upon the chuck key as part of your hand: When your hand withdraws, the key comes with it—every time. If a lathe or drill press starts up, accidentally or not, and throws a key out, someone may lose an eye; it could be you.

Files. Wherever possible, have a handle on a file, particularly when filing on the lathe, however briefly. Handles do sometimes get in the way, and then they must be removed, in which case file the tang smooth on the sides and round off the point.

Long hair. If you have long hair, wear a light cap such as painters wear when using a drill press or other machine. I once was acquainted with a man who had a lock of hair torn out by a drill press. It did not improve his appearance, nor did he recommend the experience.

Loose clothing, such as unbuttoned shirt-sleeves, can easily get caught in rotating machinery. A hanging necktie does not tear easily, and when caught in a lathe it will slam your head down with unbelievable speed. By the time the lathe stops, your appearance will be vastly changed.

Oil on the floor can cause loss of balance, and an outthrust hand can be severely damaged if it enters a running machine.

Power saws. Keep your fingers away from the blade, and always keep the floor free of sawdust; as with oil, it can cause you to slip, perhaps onto the saw.

Solvents—many of which, like acetone, are poisonous—should always be used in ventilated areas, preferably outside in an open garage or shed. Naphtha, or white gas, is a good rinse for clock movements, but it must never be brought into a cellar or basement. Being highly volatile, it can be sparked into explosion by a furnace switch or other electrical device.

Always be alert. Never take a machine for granted, or use it beyond its capacity. There is certainly no need to fear machinery, but it must be respected and used properly. Machines make excellent servants, but it is wise to remember that they also make terrible masters.

Table 7. American-British Screw Data

Screw No. or Gauge	American			British		
	TPI			BA Screw No.		
	NC	NF	OD		TPI (APX.)	OD
¼"	20	28	0.250	0	25.4	0.2362
12	24	28	0.216	1	28.2	0.2087
10	24	32	0.190	2	31.4	0.1850
8	32	36	0.164	3	34.8	0.1614
6	32	40	0.138	4	38.5	0.1417
5	40	44	0.125	5	43	0.1260
4	40	48	0.112	6	47.9	0.1102
3	48	56	0.099	7	52.9	0.0984
2	56	64	0.086	8	59.1	0.0866
1	64	72	0.073	9	65.1	0.0748
0	80	--	0.060	10	72.6	0.0669

Recommended Reading

General

Baillie, G. H. *Watchmakers and Clockmakers of the World.* London: N. A. G. Press, 1969.

An excellent reference work of 36,000 names, covering the period up to 1825.

Loomes, Brian. *White Dial Clocks.* Newton Abbot, U.K.: David & Charles, 1983.

An expanded version of *The White Dial Clock,* which contains much data on the development and dating of English longcase clocks.

Palmer, Brooks. *The Book of American Clocks.* New York: Macmillan, 1972.

———. *A Treasury of American Clocks.* New York: Macmillan, 1972.

Companion volumes which are standard works on American clocks and their makers.

Technical

Britten, F. J. *Watch and Clockmaker's Handbook, Dictionary and Guide.* London: Eyre Methuen, 1978.

An excellent alphabetical listing of traditional clockmaking and repairing methods.

Goodrich, W. L. *The Modern Clock.* Fox River Grove, Ill.: North American Watch Tool & Supply Co., 1975.

Full of sound information and interesting manufacturing details.

Norling, E. *Perspective Drawing.* Tustin, Calif.: Walter T. Foster.

An excellent book on the principles of good drawing.

Oberg, E., et al. *Machinery's Handbook.* New York: Industrial Press, 1979.

A standard work for many years; a mine of information for machinists, engineers, and craftsmen of many kinds.

Index

A
abbreviations, 15
American clocks, 22–56
American ogee, *see* ogee
anchor, 35, 65
Anglo-American, 42
anvil, 27, 129
Arabic numerals, 114
arbor, 14, 129
   assembly, 29
   escape, 17, 29
   facing off, 196
   fly, 17
   minute, 15, 17
   pivots, 24, 29
arcs, pendulum, 187
assembly of clock movement, 28–32
   box, 29

B
backlash, in gear meshing, 29, 78
Banjo 8-day weight movement, 36–37
barrels, 129
   bushing, 131
   going, 212
   hook repair, 129
   tooth replacement, 212
basic repair text, 22–32
baskets, 145, 146
   metal, perforated, 146
   wire mesh, 145
bearing, 133
beehive clock, 41, 44
bench key, 45, also *see* let-down key
bending tools, 130
bim bam, 44, 130
blemishes, 125

blocks, 168–170
bluing, 130
   of clock hands, 28
box clock, 93
Brandon clock, 59
brasswear, 121
Breguet, Abraham Louis, 17
Brocot pendulum suspension, 88, also *see* escapement
bushing
   barrels, 131–132
   extended, 136
   for pivot holes, 24–25
   making your own, 137–138
   plates, 132–137
buying a clock, 138

C
cables, removal (ogee), 24
   reassembly, 32
calendars, clock, 139
cam, slotted, 15
   synchronizing, 48
Canadian clocks, 57–61
   Pequegnat regulator no. 1 8-day weight movement, 59, 60–61
   Time clock, 58
cannon pinion, 14–15, 27, 140
cases, 116–128
   French wooden, 84
   longcases, 116
   Lyre 8-day weight movement, 33
   marble, 128
   ogee, 22
   reassembly of clock, 32
   slate, 128
   veneer, 121
   wooden, 121, 124

center gear, 15, 26, 28, 140
center shafts, 206–207
chain, 140
chainmaker, 140
chamferer, 197
chiming clock, 17, 144
chops, 66, 144
chuck, Plewes, 200
cleaning machine, 26, 144, 146
cleaning process, 24, 26, 149
click, 24, 68, 150
   inspection, 28
clock stand, 48, 150
cock, pendulum, 187
collet, 150
   friction, 166
   gear, 27
   hand, 174
column clock, 41
   and splat, 49
concepts, basic, 16
count gear, 25, 150
   assembly, 28–30
   removal, 26
count hook, 30, 151
count wheel, 25, 150, also *see* count gear
crutch, 23, 151
   settings, 159
crystal regulator clock, 85

D
dapping, 151
deadbeat, 17
depot clock, 42
dials, 106–115
   paper, 113
   removal, 23
   silvered, 113

dials. *(cont.)*
  white, 107
  wooden, 110
  zinc, 110
disassembly of clock movement, ogee, 23–26
discs, knurled, 205
dishing a barrel, 183
domes, glass, 128
Doric, 41
drilling, 151
  tables, 152, 153
drive detail, 148

E
Edwards, E. L., 14
eight-day movements, 22
English clocks, 62–83
  8-day fusee movement, 66–70
  longcase 8-day weight movement, 71–83
escape arbor, 17
escape cock, 154
escapement, 17, 154
  Brocot, 17
  deadbeat, 17
  Graham, 17
  reassembly, 31
  recoil, 17
  verge, 214

F
faking, 125
fan, 14, 164, also *see* fly
fly, 14, 164
  dismantling, 27
  English longcase, 164, 165
  forming, 164
  purpose, 27
fly arbor, 17
foliot, 65
French clocks, 84–91
  8-day spring movement, 86–91
friction collet, 166
friction spring, 167
  removal, 26
fusee dial clock, 64
fusee movement, 8-day (Eng.), 63, 66–70

G
gathering pallet, 73, 167
gear (clock), 14–16, 167
  center, 15, 26, 28, 140
  count, 15, 25–26, 28, 30, 150
  dismantling, 26
  escape, 29
  hour, 15
  meshing, 24, 28, 29
  reassembly, 29
  reduction, 15, 209
  tooth replacement, 26, 212
  tooth straightening, 26
  wooden, 213
gear expander, 166, 167
geartrain, *see* train
German clocks, 92–105
  8-day spring movement, 94–98
  Vienna regulator 8-day weight movement, 99–105
gingerbread clock, 40, 46
  wall, 43
glass cutter, 172–173
  circles, 172–173
  domes, 128
going barrel, tooth replacement, 212
gong, 44
  rod, 173–174
Gothic, round, 41, 44
Graham, *see* escapement
grandfather clock, *see* longcase
grand sonnerie, 174

H
hammer, 17
hammer lifting pins, 38
hand collet, 174
hands
  bluing, 28
  cleaning, 28
  moving backwards, 13
  reassembly, 29, 32
  removal, 23
  repair, 174
hand washer, 174
hardening
  brass, 175
  steel, 175

history of pendulum clocks, 177
hollow punches, 27, 178
hook, unlocking, 214
hour gear, 15
hour pipe, 24, 29, 178
Huygens, Christiaan, 178
Huygens system, 62, 63, 178

J
jaws, hemispherical, 204

K
key shrinker, 181
kitchen clock, 44

L
lathe, 52
let-down key, 45, 181
lifters, 17, 181
  polishing, 27–28
  replacement, 29
longcase or grandfather, 62, 65, 181
  brasswear, 121
  8-day weight movement (Eng.), 71–83
  mission 8-day weight movement (Am.), 38–39
  warps in, 116
  white-dial repairs, 107
lubrication, 17–18, 32
lyre 8-day weight movement, 33

M
mainspring, 17–18, 181–186
  winding, 46
maintaining power, 186
mandrel, 208
mantel clock, English, 64
  German, 92, 93
marble case clock, 85, 128
meshing, 28, 186
mission longcase 8-day weight movement, 38–39
minute arbor, 15, 17
motion work, 17, also *see* reduction gears

movements, clock, 22
- banjo 8-day weight (Am.), 36–37
- cleaning, inspection, repair and adjustment, 26–30
- dismantling removal, 24–26
- 8-day fusee (Eng.), 66–70
- 8-day spring (Am.), 44–48
- 8-day spring (Fr.), 86–91
- 8-day spring (G.), 94–98
- longcase 8-day weight (Eng.), 71–83
- lyre 8-day weight (Am.), 33
- mission longcase 8-day weight (Am.), 38–39
- ogee 30-hour weight (Am.), 22–32
- Pequegnat regulator no. 1 8-day weight (Can.), 60–61
- reassembly, 28–32
- Seth Thomas regulator no. 2 8-day weight (Am.), 34–35
- 30-hour weight wooden (Am.), 49–56
- Vienna regulator 8-day weight (G.), 99–105

moving a clock, 13

N
names, makers', 72
nomenclature, 14–16
number system, 14–16
numerals, Arabic, 114
nylon bushings for wooden plates, 51

O
ogee 30-hour weight movement
- background, 22
- Canadian, 58
- case, 22
- disassembly, 23–26
- repair text, 23–32

oiling, 17–18, 32, 69, 186
original (defined), 16
ormolu, 85

P
pallets, 187
- gathering, 167
- jig, 162

Pantheon clock, 58
paper dials, 113
pegging out, 28, 187
pendulum, 17, 187
- and moving clock, 13
- assembly of movement, 32

pendulum cock, 187
- adjustment, 27

Pequegnat, Arthur, 57
Pequegnat regulator no. 1 8-day weight movement, 60–61
philosophy of clock repair, 192
- makeshift methods, 25

pillar, dial, 107
pin
- hammer-lifting, 27
- removal, 24
- verge, 27
- warning, 25

pinions, 14, 193
- cannon, 14, 27, 140
- center, 15
- cut, 193
- lantern, 194
- leaf, 193
- motion, 14
- reduction, 15

pipe, hour, 178
pivots, 24–25, 194
- assembly, 29
- holes, 15, 24–25, 28
- repair, 28

plates
- bushing, 132–137
- false, 72
- front, 15
- removal, 26
- warping, 28

Plewes chuck, 200
pulleys, reassembly, 32
punches, hollow, 178

R
rack & snail, 73
ratchet, *see* clicks
recoil, *see* escapement
reduction gears, 15, 209
reduction pinion, 15
remontoire, 209

repair, clock, philosophy of, 192
- makeshift methods, 25

repairing movements, 19
repivoting, 177–179
resetting a pin, 27
rollers, 168–169
- blocks, 168–169
- mounts, 168
- shafts, 169

safety, workshop, 216
Saxon, 42
school clocks, 42–43, 122
screws, American-British screw data (table), 218
Seth Thomas regulator no. 2 8-day weight movement, 34–35
shafts, *see* center shafts
sharp Gothic, 41, 44
shellac finish, 126
shrouds, *see* pinions, lantern, collet
silvered dials, 113
slate case, 128
soldering
- silver, 209
- soft, 27, 210

spandrels, 211
split stake, 25, 211
spring
- friction, 26, 167
- hammer, 29
- pendulum suspension, 23

spring movements, 22
- 8-day (Am.), 44–48
- 8-day (Fr.), 86–91
- 8-day (G.), 94–98

springs, suspension, 23, 211
staining, 125
steeple, 41
stems, 206
stopping, 17–18
strike
- components, 54
- setting up, 30–31

strike-setting wire, assembly, 32
striking, 17

Index 223

stripping, 124
suspension spring, 23, 211

T
tambour, 40, 44
teardrop clock, 43
tempering, 177, also *see* hardening
thirty-hour movements, 22
  ogee weight (Am.), 22–32
  wooden weight (Am.), 49–56
Thomas, Seth, *see* Seth Thomas
tightening loose gear, 26
timepiece, 17, 212
tools, bending, 130
tooth replacement
  gears, 212
  going barrel, 212
  wooden gear, 213
trains, clock, 14, 17, 214
  chime, 17, 144
  ogee, 24–25
  reassembly, 30–31
  strike, 17, 25
  time, 17, 31, 144
trundles, 214

U
unlocking hook, 214

V
varnish, 127
veneer, 65, 121
verge (anchor), 65
verge escapement, 214
verges, 214, also *see* escapement
  making, 214
  reassembly, 19
verge wire, 23
Vienna regulator 8-day weight movement, 99–105
  spring, 93
  wall clocks, 123
  weight, 93

W
wall clocks, 42–43, 123
warning, 25, 214
warning pins, 25
warps
  in longcases, 116
  in plates, 28
washer, hand, 174

wax, 127
wear and tear, 18, 24
weight movements, *see* movements
weights, 17, 214
  and moving clock, 13
  ogee, 22
  reassembly, 32
  removal, 24
wheel, count, 150
wheels, 14–16, 216, also *see* gear
white dial, 107
winding mainspring, 46
wire, strike-setting, reassembly, 32
wooden cases, *see* cases
wooden dials, 110
wooden movements, 22
  30-hour weight (Am.), 49–56
wooden works clock, 49
workshop safety, 216

Z
zinc dials, 110